DESERT SURVIVAL

BY DICK AND SHARON NELSON

Illustrations by Feather Hammond
with the following exceptions:
Reptiles by Sheridan Oman
Invertebrates by Dick Nelson

CONTENTS

Introduction
1 - Water
2 - Heat
3 - Sun
4 - Problems Related to Heat, Sun, Lack of Water
5 - Clothing for Hot Weather
6 - Cold
7 - Rain, Flash Floods, Lightning
8 - Wind, Sand Storms, Dust Storms, Dust Devils
9 - Miscellaneous Hazards
10 - Food
11 - Shelter
12 - Equipment for Foot Travel
13 - Maps
14 - Lost in the Desert
15 - Always
16 - Desert Hiking
17 - Vehicle Travel
18 - Aircraft in the Desert
19 - Deserts in General
20 - Desert Plants
21 - Desert Animals
Epilogue

Copyright © 1977 by Richard C. Nelson and Sharon Nelson

International Standard Book Number: 0-915030-06-3
Library of Congress Catalog Card Number: 75-27719

Tecolote Press, Inc.
Post Office Box 217
Glenwood, New Mexico 88039
First Edition Published January, 1977
Second Edition Published January, 1978

INTRODUCTION

The purpose of this book is to provide the desert traveler with information which may help to overcome an emergency in the desert. Better yet, it may prevent the emergency from ever occurring. Surviving in the desert requires many skills which are learned through actual experience, and you won't acquire them by reading a book. For that reason, we wish to emphasize from the start that you must fortify your reading with actual experience.

In the United States alone, large areas of California, Arizona, Oregon, Nevada, Utah, Texas, New Mexico and several other states are desert. "True desert" sprawls over one fifth of the earth's surface.

There are people who have lived many years in desert communities who point out the miles and miles of "nothing" as they whiz by in their air-conditioned cars. But deserts are not just empty stretches of sand. Most deserts sustain an intriguing variety of life--trees, cacti, other vegetation and animals. The desert has no rival when it comes to smouldering sunsets, wild storms and strange animals; and there is no reason why you should not explore it and enjoy it--provided you are well prepared.

No matter what, you must develop **respect** for the desert and what it can do to you. If you venture into the desert without equipping yourself to deal with an emergency, your fate may be sealed at the outset, for the desert is not hospitable to those who drop in casually, trusting to find water if they run short. **Planning, preparation and experience are the basic requirements for desert survival.** Without them, your chances are reduced.

Planning for any desert outing will definitely help to insure that you won't be called on to use your survival skills. Never head into the desert without planning ahead. Newcomers to the desert have left their vehicles, set out to explore--and have been dead within two hours' time. Long-time residents have also paid for their failure to respect the desert, failure to plan and to anticipate its dangers as well as its beauty. The difference between a pleasant outing and a critical situation may turn out to have been a small item that you forgot to bring along, or starting out at the wrong time, or failing to recognize a potential source of water.

Survival means just that--staying alive. If you find yourself in an emergency situation, all of your energies, resources, knowledge and skill must be focused first on keeping alive, then on getting out or attracting help. Forget about everything else. Your house payment may be late, but it won't matter unless you survive. Set aside your concern for those at home who may be worrying about you. You must concentrate all your efforts, physical and mental, on resolving your problem.

Survival situations may take hours to develop (a storm may build up until heavy rains create high water between you and your destination), or a situation can develop in just a matter of seconds (a rattlesnake bite or an accidental fall). Many other factors can contribute to an emergency--heat, lack of water, ignorance, inadequate equipment, mechanical breakdown, weather and over-confidence, to name a few.

Experience seems to us to be the most important factor in desert survival. Of course, that in itself will not guarantee survival. Calm thinking, prudent decisions, planning, possibly some luck, and other factors all play a role. **Get**

your survival experience in non-emergency situations. Read books on the subject, and talk to those who have been before you. Think about what they have to say and how their advice might apply to you personally. Take classes in survival and general outdoor activities such as hiking and backpacking. Get experience every time you go out camping. Evaluate your equipment; consider how it can be improved, how you might use it in an emergency or whether it is something you really need at all. Also study ways in which you can do things without any equipment at all. Temporary shelters can be constructed without saws or axes. Water can be found without a shovel. Practice "survival" whenever you head out. Then if you ever find yourself in a real survival situation, things won't be entirely unfamiliar to you. You can think back and recall a similar situation. What got you out before? Your chances of panicking or making a foolish mistake will also be reduced with experience. Through experience, you will learn to avoid a dangerous situation or--if that fails--to cope with it.

If you get into trouble, don't reach in your pack for a book to advise you. Consider the alternatives **before** you get into trouble. Rely on your knowledge in an emergency. Try to obtain information from as many sources as possible. Consider the source, then go out in a non-emergency situation and test the information for yourself. You will be surprised how many "survival" techniques turn out to be worthless in actual practice. Don't wait until you really get into trouble to see whether a solar still is a satisfactory source of water or whether prickly pear pads will provide usable moisture. In a true survival situation, it is vital that none of your energies and resources be wasted. Remember--in a survival situation it's too late to experiment, too late to thumb through a book. **It's how well you've prepared before you head out that will determine how you handle an emergency.**

This should not be considered an instruction book or a "how to" book. Books which present you with concrete solutions to given situations are often deceiving. Some survival manuals list so many possible sources of food and water they they might cause an inexperienced individual to be careless. They create the illusion that if you fail to anticipate and provide for all your needs, you can use your book learning to live off the land. The illusion could be fatal.

We feel that "living off the land" is seldom an issue in short-term desert survival and that the effort to do so might actually hurt your chances. Desert emergencies usually terminate one way or another in a rather short time, and we decided that "living off the land" was a more sensational than practical aspect of desert survival. Except in a rare situation, your time is too valuable and your energy too vital to expend in the pursuit of kangaroo rats or ravens. We hope, instead, to offer as many practical techniques as possible, uncluttered by obscure or unproven methods. In a few areas we are non-committal. This is not because we have failed to study the concept thoroughly; it is because studies have not been made which are needed to prove whether some techniques (such as deriving water from cacti) could be more harmful than beneficial.

Our original manuscript for this book was considerably longer than the final form. As we continued researching over a two-year period, we gradually pared it down to the basics, feeling that when you've mastered them, you can go on to such things as the capture and preparation of animals for food, etc. This is an "idea" book--things to think about and test in non-emergency situations rather than exact remedies for exact situations. Remember that different people, faced with the same emergency, will respond differently. An exhilarating challenge to one person may be frightening or even fatal for another. Perhaps you recall reading in the news some years ago about a couple and their baby who became stranded in a snowstorm in the mountains in Oregon (not a desert survival case but a sad illustration of the point we're trying to make). They abandoned their car and attempted to walk out, and the wife died from exposure before rescuers found them. A man caught in the same storm was better prepared and able to keep warm with no major problems. Tragedy for one party, minor inconvenience for the other. How well you've prepared and how you respond when confronted with an unexpected situation may make the difference between living and dying. We hope this book will help you to be properly prepared for a desert emergency and suggest to you a variety of ways to either avert trouble or deal with it successfully. And don't stop here. Add to your knowledge from as many reliable sources as possible--books, veteran hikers, classes, lectures and (most of all) your own experimentation in non-emergency situations.

CHAPTER 1 - WATER

Nothing that you .ake into the desert with you is as vital to your existence as water. Yet we have encountered hikers and backpackers who seem to think it is needed only for the rehydration of freeze-dried beef Stroganoff--and we marvel that they have lived so long. Many others carry only the bare minimum required to get by, and we know of very few who actually carry enough water to see them through an emergency. Few desert drivers think to equip their vehicles with water, feeling safe in the illusion that they can always drive into a service station or a restaurant. Disaster struck a Phoenix, Arizona, family in 1970 when their car became stranded in the desert only five miles from the town of Carefree. It was August, and temperatures had soared well over 100°F. Help was delayed for about 24 hours--and a woman and four children died. There were a number of errors in judgment involved, but had there been a jeep can full of water in the car, they would not have paid so dearly.

None of the things that you carry with you into the desert will be of much use to you without water. When you are thirsty, you will find you have little or no interest in food; if you are equipped to mend a broken leg, it will do you no good if you do not have enough water to sustain life until help arrives. When you are making up your pack, consider leaving your handgun and cartridges out and substituting extra water. The water takes up about the same space and weight, but your chances of a critical need for one far exceed the chances of your needing the other. When loading your car for a desert drive or picnic, consider whether you should leave out the yard chairs and put a jeep can of water in their place.

Some people need more water than others, so **you will have to determine your personal requirements.** The first few times you go out, take more than you think you will need. Keeping in mind that you should always carry some emergency water, you can adjust to fit your needs. If you carried a lot of water back, carry less. If you had a thirsty trudge over the last mile or two, carry more. We had to come in thirsty about twice before we started tossing out bulky lunches and books and allotted the space and weight to water.

One word of caution: If you are coming in low on water, **don't be tempted to ration it.** Thirst is your body's way of warning you that it is losing water faster than you are replacing it, and you should respond to the signal as long as you have water with which to do so.

The Body's Need for Water

The human body cannot survive major fluctuations of its chemical balance or internal temperature. Your normal internal temperature is usually around 99° F, varying only slightly from one individual to the next. But an internal temperature increase of only 5° F - 7° F may be fatal. The body can compensate for great variations in external temperatures. On the other hand, its internal physiology can tolerate temperature changes only within narrow limits.

The desert can be extremely hot or extremely cold, and **you must concern yourself with maintaining an internal body temperature of about 99° F.** As the internal body temperature increases (as a result of high temperatures, exposure to the sun's rays, or exercise), the excess heat is carried to the surface blood vessels where the heat is lost from the body.

The body's primary method of cooling [and hence maintaining a 99° F internal body temperature] is through the process of sweating. Under strenuous conditions, the human body can lose up to a quart of water per hour, although the rate loss will be less under most conditions. Still, keep in mind that losing half a quart per hour would not be uncommon if you were exercising under hot desert conditions. This amounts to about half a canteen of water every hour. The individual with only one or two canteens of water is soon going to run into problems if he is unable to replenish his supply along the way.

Sweating requires water. The more you sweat, the more water your body requires. Without water, the body cannot sweat effectively enough to get rid of excess heat. The internal body temperature begins to climb, and the body chemistry begins to break down. Death may result if the process isn't reversed.

In the hot, dry desert, **sweat may not be visible on the skin;** it may evaporate before beads form. This rapid evaporation is normal but should not be confused with heat stroke (discussed later) wherein the body entirely ceases to sweat.

There are many factors which determine the amount of water your body requires. Many of these are discussed later, but they include such things as the environmental temperature; how you are dressed; the amount of wind; your age, weight and body build; the activity in which you are involved; the degree to which you are acclimatized; your physical conditioning; and others.

There have been a number of studies undertaken to determine the water requirements of people under desert conditions. The book, PHYSIOLOGY OF MAN IN THE DESERT (E. F. Adolph and Associates, 1947), includes the results of one such study. The following two charts are included to emphasize the great quantities of water people may need under desert conditions. Keep in mind that these are averages based on a number of individuals and that it is possible for some people to die in an hour or less under some desert conditions. Do not use these charts as a personal guide for determining your own water requirements; you will have to do this yourself (discussed later). Note the great influence water can have on the average figures.

NO WALKING AT ALL

Available Water per Man, U.S. Quarts

Max. Daily Shade Temp. °F	0	1	2	4	10	20
			Days of Expected Survival			
120°	2	2	2	2.5	3	4.5
110°	3	3	3.5	4	5	7
100°	5	5.5	6	7	9.5	13.5
90°	7	8	9	10.5	15	23
80°	9	10	11	13	19	29
70°	10	11	12	14	20.5	32
60°	10	11	12	14	21	32
50°	10	11	12	14.5	21	32

WALKING AT NIGHT UNTIL EXHAUSTED AND RESTING THEREAFTER

Available Water per Man, U.S. Quarts

Max. Daily Shade Temp. °F	0	1	2	4	10
			Days of Expected Survival		
120°	1	2	2	2.5	3
110°	2	2	2.5	3	3.5
100°	3	3.5	3.5	4.5	5.5
90°	5	5.5	5.5	6.5	8
80°	7	7.5	8	9.5	11.5
70°	7.5	8	9	10.5	13.5
60°	8	8.5	9	11	14
50°	8	8.5	9	11	14

(From PHYSIOLOGY OF MAN IN THE DESERT, E. F. Adolph and Associates, 1947)

The desert traveler is going to ask: "How much water will I need if I am going on an all-day hike in the hot desert?" Unfortunately, there is no set answer. Under all desert conditions, we suggest that one gallon of water per person per day be considered the absolute minimum. Keep in mind that this would not be nearly enough in many situations.

One practical way to determine your water needs under various conditions is to start out with more water than you are sure you will need. For a day hike, you might take six quarts of water. Keep a notebook; record temperatures, terrain, distances and times; record how much water you have left. The next time out, you may find that you can get by with less water--but always have some extra in case an emergency develops. The idea is to **work down to your water requirements rather than up.** From our experience, we can judge quite closely how much water we will each need under various conditions; thus we are sure to have sufficient water without carrying much excess.

Try to avoid formulas for determining your water needs. We often see figures such as: "You can hike ten miles with one gallon of water." Under certain conditions, that one gallon might take you a mile; under other conditions, it might take you thirty miles. **Only practical experience will enable you to determine your water needs.**

Weight is an important factor when considering how much water to carry. One gallon weighs just over eight pounds. To this must be added the weight of the container. At some point, weight will become the limiting factor which will determine how much water you can carry with you. While you may be capable of carrying three gallons quite easily (total weight, including containers, of over 25 pounds), keep in mind that the weight of the water alone will increase stress on the body and result in a greater heat production which will require more sweating and more water replenishment. We have never carried more than two gallons per person. Whenever our demands were greater, we planned on refilling at definite sources or limiting our hike. A good rule of thumb is that when your water is not quite half gone, your trip should be at least half completed.

We can think of no circumstances where you should head out into the desert on foot without any water at all. Even in winter, you will need some. If you are positive that there is water at your destination, you will still want some water with you in case an emergency arises enroute.

Always carry plenty of water for your anticipated needs plus extra for emergencies. As you gain experience in desert travel, you will probably have a tendency to carry more water. Running out of water is, at least, no fun; at worst, it can be fatal.

Thirst is not always the best indication of your body's need for water. **You can be dehydrating quite rapidly and not feel terribly thirsty.** Whenever you are out in the hot desert, drink at regular intervals rather than wait until you're thirsty. When your body becomes short on water, your brain's ability to recognize trouble and signal you through thirst may be impaired. Your overall physical and mental capacities may be impaired as well. If the weather is hot, we drink water regularly and encourage other members of the group to do the same.

The selection of water containers should not be a casual process. Don't pick up the first one you see at the local discount house; many of these are flimsy or unsuitable and could let you down at a critical time.

After some trials and errors, we've come to prefer polyethylene canteens with about a one-quart capacity. They have a double closure which consists of a plastic plug plus a cap which screws down over it. The cap and plug are attached to the top of the canteen to insure against loss. This type of canteen is available at good outdoor stores which carry backpacking supplies.

You will have to experiment with various types of canteens until you find the size and type best for you. Some kinds give the water an unpleasant taste. Others are poorly made and last only a short time. (Some of ours are ten years old now.) Some canteens have a cloth covering on the outside which helps to cool the contents through evaporative cooling. We don't use them because of their heavier weight.

The night before heading out, we fill our canteens (leaving a little room for expansion) and place all except one per person in the freezer. Some kinds of canteens will split if you freeze them, so you may have to experiment with different brands to find a tough one. Keep one canteen refrigerated but unfrozen so that it will be cool and drinkable at the start. Using this method, you can have cool water well into the day, even on a backpack. (Avoid drinking cold water rapidly.)

Conserving Water

The best way to conserve the water you have with you is to avoid situations where you sweat. Never ration water if you are really thirsty because the thirst indicates that your body needs water. The water in your body will help maintain the 99°F internal temperature; the water in your canteen will not. Many people think they can stretch their water supply by sipping small amounts even though they are sweating heavily and are thirsty. This is not the case. If you are dehydrated, your body needs water immediately, not at some future time. By trying to sip, you may impair your chances of traveling far enough and thinking clearly enough to get out or find a water source.

Again, the need for water cannot be over emphasized. If you are sweating, if it is hot, or if you are thirsty, be sure to drink plenty of water. Also be sure that your salt intake is sufficient, especially if you are sweating profusely.

There have been many cases recorded where people have been found dead while still having plenty of water in their canteens. While there are many possible explanations for these deaths, it seems that at least some of the victims simply tried to ration their water rather than drink when they needed it. Get the water into your body. Drink while you can, and make every effort to conserve the water after it is in your body by doing everything possible to reduce the need to sweat. There are many ways to cut down on sweating, and we go into them at the end of Chapter 4.

Effects of Lack of Water

This important aspect of desert survival is covered in Chapter 4 where we have also discussed the effects of heat and sun on the human body. It is often a combination of these and other factors which results in serious problems.

Carrying Water

We have already mentioned the polyethylene containers we carry with us on daypacks and backpacks. Ours hold about one quart and have an attached plug and screw-on cap. They are unbreakable, and we have been able to freeze their contents without having them break or split. They come in assorted sizes and shapes for ease of distribution in your pack.

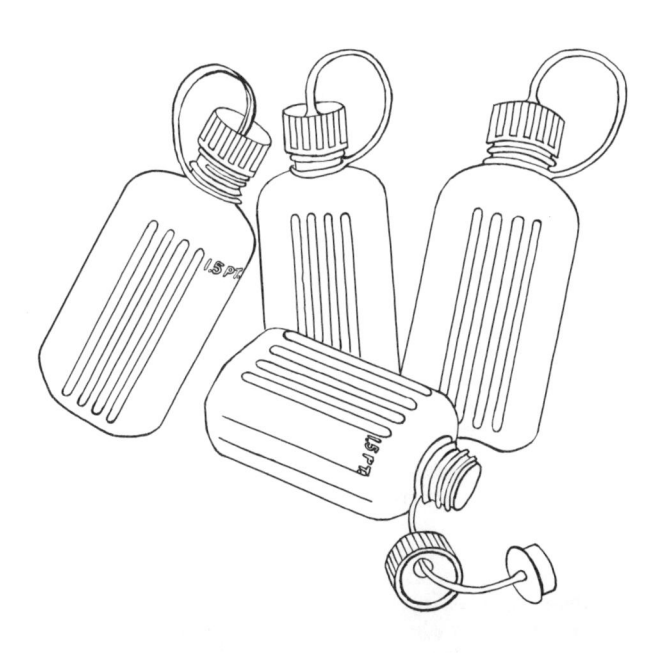

    We feel it is safer to carry several canteens rather than a single large water container. They are easier to fit into your pack, and the weight can be evenly distributed. More important, a leaky container or a spill won't threaten your entire supply of water. When you empty a container, place water purification tablets in it, and collect water at the first source you come to; thus the water will have been purified should you have to resort to drinking it. We like to flavor at least one canteen with lemonade or a similar drink mix to provide variety.

    For vehicle travel, we have heavy-duty polyethylene containers which hold two and a half gallons each--a convenient size for packing and not so heavy when full that a small person can't handle them. You will have to do some searching to find good ones as many that are on the market are not durable enough. For longer trips, we usually carry four of these containers plus a plastic cooler which holds a large block of ice. (The block is better than ice cubes or crushed ice for keeping water colder longer.)

    Label your water containers to avoid confusing them with other containers that might contain fuel, etc. Keep track of which are full and which are empty. There is nothing more frustrating than to reach into your pack for the last full canteen on a hot day only to discover that you have already emptied it.

Always use soap and water to wash out a new water container before using it. Check for unpleasant odors and tastes before heading out with it. We have purchased several that gave the contents a strong "plastic" taste.

Avoid flimsy water containers or any which have parts that can rust. Collapsible containers give you that one advantage but are usually too thin-walled to be reliable. Water bags scratch vehicles, get dirty and may fall off unnoticed.

Always fill canteens and water containers with fresh water before heading out. When you return home, empty them and leave them uncapped to help prevent bacteria and fungi from growing.

The foot traveler should not overlook his stomach as an excellent reservoir for water. Drink plenty of water from the supply carried in your vehicle before you start to hike. We try to drink as much as a quart if possible, despite the fact that we may not feel thirsty. This gives us an extra quart each without taking up the space or weight of a canteen. Alcoholic beverages hasten dehydration, and soft drinks are impractical. It's wisest to carry water, with one possible exception. Some of the new beverages designed for athletes contain salts, sugars and water. You might find some of these products helpful in keeping the necessary salts in your system.

Finding Water

Those desert travelers who depend on finding water rather than carrying plenty with them are asking for trouble. **Always carry with you sufficient water to meet your needs.**

Potable water is not available in many desert areas. It's true that you may be able to find water in certain areas, but natural sources are usually few, situated many miles apart and may be concealed even from the trained eye. Springs marked on maps may be seasonal only. Even so-called "lakes" may be dry or have highly alkaline water. Below we have described some of the places and ways to find water in desert areas. **Keep in mind that these are only possibilities, not certainties.** Any water you do find should be treated. It may be contaminated even though no human habitations are near. In desert environments where water sources are small and far apart, the wildlife alone may pollute the water so that it is unsafe to drink unless treated. (See next section, "Treating Water for Drinking.") There may be some possible water sources not mentioned below, depending on your location and situation.

1. **Stock ponds, stock tanks, windmills:** Ranches are scattered throughout the desert Southwest and are found in many other deserts throughout the world. Unfortunately, not all have good water (though a fouled source could save your life in an emergency, and you could deal with the consequences later). Before starting toward a distant windmill on a hot day, consider how likely you are to find water there; some windmills haven't pumped a drop in decades. If possible, treat all water with purification tablets.

2. **Ponds and streams:** Most southwestern streams are dry. Check your topo maps for locations of live streams and wet ponds. Many of these may flow underground for a distance only to surface farther downstream. Others may be strictly seasonal.

3. **Tinajas:** In many rugged desert canyons there are hollowed-out depressions in rocks. These depressions are known as "tinajas" and may contain water. They vary in size from a couple of inches wide to large ones

many feet across. Some of these hold water for a few days after a rain, and others hold water the year around. They may be important water sources for animals such as desert bighorn sheep, though cattle and burros may foul the more accessible ones. If algae covers the surface, you may be able to clear it away by pulling a T-shirt across the surface. Many hikers carry a fine mesh cloth with them just to filter larger particles from such water sources. Also, allowing the water to settle in a canteen for a while will enable you to pour water off into another canteen, leaving the sediment behind.

4. **Seeps, springs, cliffs:** Along the bases of many cliffs, especially in mountainous areas, you may find small, moist areas where water seeps out. Some springs are also marked on topo maps, though many are only seasonal, and some are dry and no longer a source of water. If you find a place where a small amount of water can collect, dig out a small depression and allow it to fill. Take a piece of rubber or plastic tubing from your emergency kit (described later) and suck out the water. You may have to place a fine piece of cloth over one end to filter debris. Sometimes it is possible to set up a siphon so that water will be collecting in your canteen while you are working on other projects such as gathering firewood.

5. **Green areas:** Again and again we are told that all you need to do to find water in the desert is to look for an isolated green area. We have consistently checked these out and find that very few of them actually have surface water. In fact, the water beneath a clump of green mesquites may be fifty or more feet below the surface. Some trees are more indicative of water close to the surface than others. Cottonwoods, sycamores and seepwillows are among those. In addition, a green thicket of vegetation can provide excellent shade until a cooler time of the day. Green areas high on hillsides also suggest surface water but may be inaccessible to a person already suffering from dehydration.

6. **Digging:** If you find a moist area on the surface of a sandy wash, it may be easy to dig down a short distance for water. Allow it to collect and settle in a dug-out depression, then scoop it out or use your rubber tube to suck it up. Sometimes it may be necessary to dig down many feet to reach water even though the surface is moist. Check carefully as you hike along the edge of a wash for the most likely spot. Such a spot would often be where the wash makes a curve along a cliff.

7. **Barrel cacti:** In the minds of a great many people, the barrel cactus seems to loom as a dependable source of water. Reports by early expeditions indicate use of barrel cacti by Indians; however, we consider the barrel cactus controversial. We know of no comprehensive study which has determined the safety and reliability of the plant as a source of potable liquid. Barrel cacti are common in many desert areas (and non-existent in many others). Some barrel cacti reach four or more feet in height. They are covered with many large, robust spines and have a tough outer covering. It is difficult to cut the cactus open with just a handknife; a machete or saw is almost required to do the job easily. Inside is a pulp and liquid. Some people have reported consuming large amounts of this liquid (it may be necessary to mash and squeeze the pulp to get any quantity of fluid); others have reported becoming nauseated and throwing up after drinking it. Some barrel cacti pulp/liquid is almost tasteless; some is very acrid and bitter. Certainly a dehydrated person without equipment would find it difficult to open a barrel cactus. Even then, there is the possibility that he may become sick on the contents.

Our experimentation with barrel cacti has been limited. The cacti we tried weren't palatable and didn't lend themselves to experimentation on a large scale. A comprehensive, scientific-medical study is needed which involves large numbers of people in various stages of dehydration (plus control groups). An analysis needs to be made of the chemical contents of the pulp and liquid to determine whether there is a seasonal variation, a variation between species of barrel cacti, etc. From our experience, just getting at the pulp is a major chore and a questionable expenditure of energy for a dehydrated individual.

8. **Solar still:** The solar still seems to be a popular topic with the authors of survival books. Our experience has indicated that there would be very few instances where we would even consider the use of such a device.

Basically, a solar still involves a hole dug in the sand or soil (preferably moist) with a sheet of transparent plastic covering the hole. Sunlight passes through the plastic, warming the soil and the air between the sheet and the ground. This results in evaporation of water from the ground. The water condenses on the sheet of plastic and runs down to a point where it drips into a collecting vessel such as a jar or canteen. The apparatus works best in full sunlight, and the more moist the sand, the more water you are likely to obtain.

There are several problems which we have encountered with solar stills. First, it requires a fair-sized hole (say three feet across and a foot and a half deep). Digging this hole takes some time and considerable effort. If we had already been in trouble, just digging some of the holes would have been enough to increase dehydration and result in an increase in body temperature. In a real survival situation, death could have resulted. If you succeed in the excavation of the hole, you must have the necessary sheet of plastic with you, and the plastic sheet is useless if you tear it or punch a hole in it. Then, even if the still is properly functioning, it takes several hours to get water, and the yields may be very small. These can vary from less than a pint (hardly justification for setting up a solar still) to three or four pints under good conditions. Even then, you would need several operating stills to get a minimal

The solar still does produce water, but we suggest that you set one up in a non-emergency situation and evaluate the results before depending on one as a ready source of liquids in any desert survival situation.

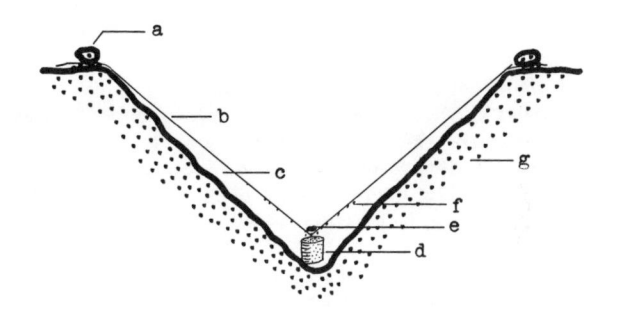

a - rocks and sand to hold down edge of sheet
b - sheet of thin plastic
c - humid air space between sand and sheet
d - container to catch dripping water
e - smooth rock to hold plastic sheet down
f - droplets of water gathering on underside of sheet
g - moist soil or sand

A cross-section view of a "typical" solar still. Approximate dimensions 3-4 feet across, about 2 feet deep. A rubber tube may be added going from the container [d] to the outside.

amount of drinking water. Because of the energy required to excavate the holes and the fact that the thin plastic may not lend itself to re-use, you would probably be forced to remain in the area where the stills are located.

Rather than come out and say to forget the still, we suggest that you get the materials and set one up in a non-emergency situation--perhaps in your yard or on your next camping trip. Keep notes on the time and effort involved and the water yields. Also keep in mind that you will be setting it up under different conditions in an emergency. Try digging the hole with a piece of wood or other material that you would have to use in a real emergency. Then decide for yourself whether or not to carry solar still construction materials around with you.

In constructing a basic still, dimensions, etc., will vary. Find a moist spot such as a sandy wash where digging will be easy. Dig a hole three to four feet in diameter and about two feet deep, and center a canteen or other collecting vessel in the bottom. If possible, place a plastic or rubber siphon tube in the container and run it out of the hole. This will enable you to draw water out without disturbing the humid atmosphere you hope to create. Place a sheet of transparent plastic over the hole in such a way that the center of the plastic will hang downward above the mouth of your container. (Dupont's "Tedlar" is often suggested because it has a somewhat roughened surface which allows water droplets to collect more easily. Some brands of plastic release the drops of water before they can run down into the collector.) If one side of the sheet is slightly roughened, the rough side should be on the inside (down). Place a smooth, rounded stone in the center of the plastic to form a point just above the mouth of your collecting vessel. It may be necessary to cover the stone with some leaves or other material to help prevent it from absorbing too much heat and melting through the plastic. Check to be sure that no part of the plastic sheet is touching the ground except around the upper, outer edge of the hole. Use rocks and sand to close this outer perimeter, being careful to seal it completely. If the sun is shining, you should soon be able to see the plastic fogging up as moisture begins to condense on the inside. It will probably take a few hours to really get going. Do not disturb the still unless absolutely necessary. It may continue to function into the night, but it will definitely do best in direct sunlight.

There are various modifications of the basic still. One method involves chopping up moist vegetation (such as the pads of prickly pear cactus) and placing it inside the still to provide an additional moisture source. Keep in mind that this would involve additional effort and that, in a real emergency in the hot desert, it is imperative that you avoid any unnecessary sweating if possible. Another modification is a somewhat different still which does not involve digging. It makes use of chopped vegetation placed in a plastic bag. We are not going into any details on this type of still because we believe there is an outside possibility that, if the wrong type of vegetation were used, it could have harmful effects. Possibly this technique should be tested in conjunction with a study of the pulp of barrel cacti.

The nature of a solar still requires that you stay in its immediate area. There are a few hypothetical situations which we can think of (such as a downed aircraft in a remote area) where a person might be able to extablish a number of stills and obtain sufficient water for survival. Lizards and snakes might sometimes fall into the still and could theoretically provide a source of emergency food.

9. **Dew:** Dew is not likely to provide water in any quantity, yet it is important in a survival situation to consider all potential water sources. Dew collects on the surface of rocks, vehicles and aircraft early in the morning. If you clean the surface the night before, it may be possible to obtain some potable water by using a piece of cloth as a sponge, then squeezing the water out into a container.

10. **Rain:** If you see a rainstorm approaching, set up your tent, rain fly, poncho, tarp (anything feasible) in such a way that you can collect rainwater. Set out all your empty canteens and containers, and try to have everything ready in advance because desert showers are often over as suddenly as they begin. Immediately after a rainstorm, check washes, holes in rocks and similar depressions for water.

11. **Ranches:** There have been instances where people suffering from dehydration have walked right past occupied ranch houses. Check with binoculars to see whether or not a distant ranch house appears occupied before risking additional exposure to the elements. It may have been long abandoned.

12. **Caches:** For the cross-country traveler on an extended jaunt, caches of water can be a very satisfactory method of helping to insure a water supply. Never depend entirely on them as you may not be able to locate the exact site, you may decide on a different route, or rodents or people may tamper with them. Glass containers are probably more satisfactory because they are less likely to be chewed through by rodents. Make a map showing the exact locations, and try to obscure the sites to prevent curious individuals from discovering them. An additional advantage of caching water is that it enables you to make note of landmarks, terrain, etc., along portions of your proposed route before actually starting out. Those rugged individuals who hike across Death Valley and undertake other arduous journeys often set up caches. Another way to extend your water supply is to have someone meet you at predetermined spots along the route. This provides a safeguard in that someone will be looking for you or reporting you lost if you don't show up at a predetermined time.

13. **Game and cattle trails:** Both wild and domestic animals must water regularly, and it is sometimes possible for an experienced individual to interpret their trails and follow them to water. Cattle come in to water in the mornings and again in the evenings. Because cattle move more slowly and more obviously than most wild animals, it is fairly easy for the foot traveler to follow them (or their dust) to a water source, especially in the evening. It is important to determine that the cattle are not leaving water but heading toward it. Try to observe the animals from some distance so as not to disturb their normal behavior pattern. Study trails to see which way they are headed. A sensible approach to game and cattle trails would be to watch them from a distance in the early morning or late afternoon to see which direction the animals take.

14. **Birds:** Birds do fly toward water at certain times, but because of their great mobility and the distances they can cover in a short time, they are more difficult to follow. Still, make note of their general flight directions, especially early and late in the day.

From time to time, water can be found using other methods and techniques and in other locations. **Planning and preparing by having plenty of water with you is really the best guarantee you have. To go out and find drinkable water, especially under emergency or survival conditions, is often** very difficult or even impossible in many desert environments. Remember that a person who is already dehydrated will no longer think as clearly as he or she normally would and that travel may not be easy to undertake. **When deciding whether to check out a possible source of water, you must weigh the probability of actually obtaining water against the effort required, the amount of dehydration you're risking, your condition, location and many other factors.** Only you can decide whether or not it is a good idea to climb a rugged hill, for example, to check a green copse for water. Experience will be your best teacher. No book can give you a pat answer for every situation that might arise.

Treating Water for Drinking

We almost always treat all water obtained from outside sources, including water from springs and running streams. Bacterial contamination may result from man's activities upstream, from wildlife and from other sources. The only instance where we might not take the time to treat water would be in a real emergency where water in the body would be more important than the possible consequences of bacterial contamination. In many areas, even campground water should be treated.

There are two basic field methods of treating water to kill bacteria. One is to bring the water to a full boil for at least ten minutes. This is usually impractical as you would have to build a fire (undesirable in hot weather), perhaps have to collect fuel (additional stress) and allow the water to cool (valuable time lost).

The method we use is the addition of water purification tablets to the water. These tablets are available at most drugstores and at many outdoor stores. Follow directions on the label as use may vary from one brand to another. Do not try to short-cut the time required for treatment. The tablets may impart a slightly unpleasant taste to the water which can sometimes be reduced by shaking the treated water and leaving the cap off for a while. We like to carry flavored drink mixes to add to treated water. When it appears that we might run low on water and we come to a water source (even if it's a questionable one), we fill our canteens and add tablets. Then if it should be necessary to drink this water, it has been properly treated by the time it's needed.

Bad Water

In times past, the hapless driver could drain his radiator and get drinkable water, albeit a little rusty. Today we wouldn't risk the radiator as a source of drinking water because of coolants, various rust inhibitors and additives. You'd be risking a fatal potion.

The term "Bad Water" reminds us of a location in Death Valley by the same name. It is about 280 feet below sea level, temperatures are high, and the shallow pools contain highly mineralized water. Some algae and a few species of insects inhabit it, but it is not drinkable. We have had no personal experience with bad or poisonous water. There is no easy, positive way to determine in the field whether or not certain water is poisonous, but we would never drink from a source where no algae was growing in the water and no insects were swimming around. Check the edges for dry minerals which might indicate an alkaline condition. While such water might not be poisonous in the usual sense, it could easily result in gastric disturbances and diarhhea--which would be serious trouble for an individual who was already dehydrated. Posted signs warning of bad or contaminated water have a tendency to disappear. Note whether or not animals have been using the source. Again, this is not a fool-proof assurance of safe water, but it will help to eliminate at least some questionable sources. If there is any doubt in your mind, and if other water is available, then the best policy would be to pass up a questionable source of water.

CHAPTER 2 - HEAT

Heat is usually expressed in degrees Fahrenheit (F) or Celcius (C), and it can critically effect the human body. As mentioned previously, human physiology functions best at an internal temperature around 99°F, and even a slight deviation can cause serious problems. Heat can be an ally when environmental temperatures are low; it can be a deadly adversary in the hot desert. **The effects of heat often combine with other imbalances such as a shortage of water, insufficient intake of salts, exposure to sun and wind, exercise, stress and other factors.**

If you are visiting a desert area for the first time on a hot summer day, the first step you take from the airplane door is like walking into a blast furnace. The longer you live in the desert, the less you seem to notice the heat; but it has the same deadly potential for native or newcomer.

Death Valley is consistently one of the hottest places in the world for part of the year. It is a beautiful expanse of true desert--brittle and cold in the winter and shimmering in the heat of summer. Portions of it are actually below sea level. To give you some idea of what summer temperatures can reach there, we have listed the minimum and maximum readings for a seven-day period in June and July of 1973. (This was considered to be a hot spell even for Death Valley.) All readings were taken in the shade and are in degrees F.

	0600	Maximum Temperature
June 25	--	118
26	87	123
27	98	125
28	90	122
29	94	121
30	93	118
July 1	88	112

Note the very high minimum temperatures. At no time during this period would strenuous activity have been advisable. Add to these temperatures direct sunlight and additional heat from the ground, and the potential stress on the human body would be tremendous.

Surprisingly few people have died in Death Valley from problems related to heat, sun and lack of water. While thousands of visitors pour into the area each year, most do not stray from the paved roads. Just the place names--"Death Valley," "Funeral Mountains" and "Furnace Creek"--seem to instill caution in most visitors. Roads are well patrolled by rangers, and most people heed the warnings and advice dispensed by bulletin boards, signs and leaflets.

Even within a localized area, there can be a tremendous variation in temperatures. The difference between readings taken in the shade and in the sun may often be many degrees F. Several feet down in a rodent burrow, the temperature might be a comfortable 75°F while the surface sand is too hot to touch.

Temperatures even a short distance off the ground are often many degrees cooler than surface temperatures. Therefore, if you must stay in a particular location for some reason, try to at least sit on something above the ground--a branch, a stone, a pile of rocks in the shade, etc. **If the air is hot, always rest in the shade if possible, and try to avoid exposure to wind if dehydration is a problem.**

Get yourself a thermometer in a plastic or metal case, and take it along next time you hike or camp. Experiment by placing it in different spots to see what types of areas are coolest. Large masses of rocks radiate a great deal of heat which you would not want to be absorbing in a hot desert survival situation.

If you begin to feel weak or think you might collapse, get into the shade as soon as possible. Get any victim into the shade and off the hot ground. One potential problem with solo hiking is the possibility of fainting for falling on hot ground exposed to the sun. If the victim doesn't revive quickly and if temperatures are high, the body may lie in the heat until death occurs.

Most people can tolerate even very high air temperatures for short periods of time. It has been demonstrated with military personnel and other groups that **acclimatization helps reduce heat problems.** Proper pacing, experience, suitable clothing, limitation of activities to cooler times, avoidance of stressful situations, plenty of water and salt, planning and preparation will all help to greatly reduce your risks in hot desert country.

Record high and low air temperatures should be thought-provoking if you're planning a desert trip. We include records for two areas here. Parts of Death Valley are below sea level; Tucson, Arizona, averages a little over 2,000 feet in elevation. Death Valley is in the Mohave Desert; Tucson is in the Sonoran Desert. All recordings were made in the shade and are in degrees F.

	Death Valley		Tucson	
	High	Low	High	Low
January	87	15	87	16
February	91	27	92	20
March	101	30	92	20
April	109	35	102	27
May	120	42	107	38
June	125	49	111	47
July	134	52	111	63
August	126	65	109	61
September	120	41	107	44
October	110	32	101	26
November	97	24	90	24
December	86	19	84	16

While these figures may fit more into the "interest" category, there are several things which should be pointed out. Note that very high temperatures may occur in seasons other than summer. Even in March and October, the thermometer may rise to 100°F or more. Record lows indicate that even the "low" desert can be bitterly cold, especially if the wind is blowing (discussed in Chapter 6).

High ground temperatures can pose serious problems for hikers and backpackers. Even the well-prepared individual may find these temperatures too hot for comfort, and blistering and other foot problems may occur. High ground temperatures can also result in an increase of heat absorption by the body. Temperatures on the ground are often 30 degrees higher than air temperatures in the shade, and readings of 50 or even 60 degrees higher are not rare in some areas. The personnel of the National Park Service in Death Valley have recorded some astounding ground temperatures. In 1974 a reading of 186° F was recorded on the ground and in 1975, a reading of 182° F. In 1972 a reading of 201° F was recorded. This is approaching the boiling point of water! With high ground temperatures in mind, carefully consider the insulative value of your footwear in addition to such basic considerations as traction, durability, weight and support.

CHAPTER 3 - SUN

In warm or hot weather, the desert traveler must avoid exposing the skin to the direct rays of the sun. Ultraviolet rays cause burning and, on a long-term basis, can cause skin cancers. **Exposure of your skin to the direct sun will result in the body absorbing large amounts of heat.** This, combined with other factors such as dehydration, stressful exercise, illness, etc., can brew up a fatal combination. These effects are covered in more detail later, but remember that it is often a combination of factors which get people into real trouble, not just a single occurrence.

Desert air is often clear, so the sun's rays may be more intense than you're used to. Desert soils in some areas are very light in color, reflecting additional light onto the body. Large bodies of water in some desert areas pose glare problems for boaters. Whatever your activity in the desert, avoid direct exposure to the sun's rays.

Sunburn

Sunburn may occur very rapidly in desert areas. Usually it results in little more than discomfort for a day or two; but it is potentially very dangerous. Sunburn impairs the ability of the sweat glands to function normally. The internal body temperature of an individual in a survival situation, possibly dehydrated and exposed to high environmental temperatures, could easily build up to a dangerous or fatal level. All areas of exposed skin will absorb heat from the sun's rays, but the back of the neck is especially vulnerable to heat. Major blood vessels going to the brain are located here and can quickly absorb enough heat to warm the blood going to the brain. Always protect the back of the neck from direct sun. A bandana or wide-brimmed hat would usually be effective. If nothing else, tie a spare T-shirt or other material around your neck. Clothing for desert travel is described later, but be sure to wear a long-sleeved shirt, long trousers and a hat. Typical, flowing Arab garb (which may look cumbersome and hot) evolved from the need to expose as little of the skin to the sun as possible.

Sunblindness

Sunblindness is not too common in desert areas, but it can occur if you are careless enough to over-expose your eyes to sunlight--and it can be very serious when it does occur. Sunglasses can help reduce the sun's glare. We know some people who always wear sunglasses on the desert and others who never do. Even regular eyeglasses can help to provide some protection from blowing sand and thorny twigs.

Sunblindness is more likely to occur on a lake or in areas where the soil is very light. It should be considered serious, and immediate first-aid steps should be taken. Treatment by a physician should be arranged as soon as possible. It may not be possible for the victim to continue on for a day or two.

If there is excessive glare, dark sunglasses and a wide-brimmed hat will help. Even regular eyeglasses can be contrived to provide makeshift protection. Take two pieces of adhesive tape from your first-aid kit and place them on the lenses, leaving a thin slit between them. You will have to apply them carefully in order to situate the slit in your direct line of vision. Tape can also be placed along the bows of the glasses to provide a lateral sun screen. Pieces of cardboard or plastic could also be used to make emergency sunglasses. Cut horizontal slits in the pieces of material, and tie them on with string, shoe laces or whatever your pack provides. Even a spare T-shirt (if relatively thin) can be placed over the head to provide protection.

The best way to prevent sunblindness is to avoid long exposure to the sunlight on bright days. Restrict outdoor activities to times when glare is not a problem. Again, the experienced desert traveler--by protecting the skin from direct exposure to the sun's rays and by protecting his eyes--avoids a multitude of problems which may trouble the novice.

CHAPTER 4 - PROBLEMS RELATED TO HEAT, SUN, LACK OF WATER

This is not a first-aid book. We want to make you aware of as many alternatives as possible should you be faced with a desert emergency. One of the best weapons for desert survival is a thorough knowledge of first aid which includes diagnosis and treatment of a number of symptoms. We do include some brief descriptions of medical problems, but these are not technical enough to be used for diagnosis--nor are they intended to be. They are included only to make you aware of certain problems. We strongly recommend that you acquire your knowledge of first-aid treatment and diagnosis through the American Red Cross or some other organization.

It doesn't take many hours to complete a basic first-aid course, and the chances are good that if you travel much you will use the knowledge and skills you acquire. Try to recognize medical problems before they become serious. Better yet, try to prevent them entirely through prudent activity and careful planning. In addition to learning the basics such as treatments for sprains, breaks and bleeding, all desert travelers should be thoroughly familiar with the diagnosis and first-aid treatment of sunblindness, sunburn, bites of venomous snakes and insects and other arthropods, heat cramps, heat exhaustion and heat stroke.

The human body consists of billions of tiny cells, each of which can function and survive only within very narrow temperature and chemical limits. All are organized to work together in one way or another, and excess heat, cold, sunlight, various activities and lack of water and salts can all upset this delicate balance.

Homo sapiens inhabits environments ranging from the cold artic to steaming jungles and parched deserts. Despite a tremendous variation in environmental temperatures, people actually live within a much narrower range of environmental temperatures regulated by shelters, clothing, heating and air conditioning.

The body's main method of cooling is through the process of sweating. If it is hot out or if the body temperature is going up due to exercise, there is a pooling of blood in the vessels near the surface of the skin. This process brings warmer blood to the surface where sweat is formed to evaporate the excess heat into the air. As a result of this pooling of blood near the surface, however, there is a reduction of blood volume in the inner body. A reduced flow of blood is available to supply vital organs and muscles with the nutrients, oxygen and removal of waste products that they require for efficient functioning. As a result, a person sweating heavily in a hot environment will feel tired and suffer a reduction in overall body efficiency. With less blood in the inner body, the heart will have to pump harder to provide organs with blood, and this results in additional stress. Keep in mind that this additional stress may occur at a time when you are already under a considerable strain. It might occur near the end of a long hike when you have lost the route, for example. It isn't too hard to understand how some people can get into trouble on the hot desert rather quickly.

Dehydration occurs when there is an insufficient intake of water [and sometimes salts] to compensate for losses. Whenever you exercise, sweating will occur, and there will be some water lost from the body. This is normal, providing that the water is soon replaced and the amount of dehydration does not become excessive. Remember that due to pooling of the blood and other factors, the more dehydration occurs, the greater the reduction in overall body efficiency. If nothing else, simple thirst resulting from slight dehydration will reduce the enjoyment of any outing.

There are many symptoms of dehydration. Being thirsty is one, but you can also be dehydrating without becoming very thirsty. When the amount of dehydration increases, "cottonmouth" may occur. This is simply a term to describe a very dry mouth. If you have ever experienced it, it's uncomfortable. Victims of dehydration may develop speech problems such as a slowness of speech or inability to speak clearly in a normal manner. There may well be a tired or lazy feeling, a reduction in muscular efficiency, a slowness in the hiking pace and other body movements. A darkening of the urine is a good indication that the body is suffering from dehydration to some degree. Victims often suffer from headaches and may feel dizzy. There are additional symptoms with which you should become familiar. Symptoms will vary somewhat with the circumstances, the individual and the severity of dehydration.

Dehydration can easily sneak up on you before you realize that a problem is developing. As the extent of dehydration increases, your ability to recognize it is reduced. In fact, if water is not readily available, your ability to do anything about the situation is much reduced.

Prevention is the best method for dealing with dehydration. Don't wait for symptoms to appear. Drink plenty of water at regular intervals. Be sure to eat well and get plenty of salt. Avoid stressful situations which involve excess sweating.

In Chapter 3, we discussed sunburn and sunblindness. Besides dehydration itself, there are three other major problems associated with heat, sun and lack of water/salts which commonly occur in the hot desert. Keep in mind that dehydration can contribute to or be a part of all three of these problems which are heat cramps, heat exhaustion and heat stroke.

Heat Cramps

Heat cramps quite often develop when there is a loss of salts from the body. This is likely to occur after a period of heavy sweating when both water and salts have been lost without being replenished. Salts are necessary to maintain the proper chemical balance of the body cells. Heat cramps may occur in arid desert areas or in hot, humid areas.

A victim of heat cramps may be hiking along the trail one minute and be face down in the dust the next--although the depletion of salts and a chemical imbalance will have been building up for some time. Symptoms typically include painful cramps in various body muscles.

It is important to remember that heat cramps may occur even when there has been sufficient consumption of water if salt is not being replaced as well. Heat cramps can usually be prevented by making certain that you consume plenty of salts along with your water and by avoiding stressful situations which involve heavy sweating. After administration of first aid, the victim should rest thoroughly and try to avoid stressful situations for at least a few days.

We take an occasional salt tablet during hot weather, and we make a point of taking along salty foods to nibble on while hiking or driving in hot weather.

Consider taking potato chips, pretzels, jerky, peanuts, cashews or sunflower seeds for snacks and salting your food more heavily than you normally would. Some desert travelers and people who work out in the hot desert take salt tablets as a matter of course, others don't. Some studies have shown that for heavy work in the hot desert, the salt from foods alone may not be sufficient to prevent heat cramps. Under hot, strenuous conditions, a salt supplement is required. Salt tablets are less likely to cause gastric disturbances if they are dissolved in water or eaten with food. They should be considered a major component of any good first-aid kit.

As with water, the need for salt will vary with the situation and the individual. Ask your doctor for advice, and keep notes on your individual needs. Suggested dosages are usually given on the containers. Salt tablets are usually available at drugstores and outdoor stores.

Salt taken without sufficient liquid intake will hasten dehydration; **don't take salt if you haven't enough liquids.**

We have done some experimenting with the commercial drinks designed for athletes. We find that they do help prevent fatigue in the desert, and we now fill one or two canteens with these drinks instead of water--but we still carry several canteens of water.

Heat Exhaustion

When the body is under stress (heat, etc.), the blood has a tendency to pool near the surface of the skin, as we've discussed before. When the vital organs and muscles are deprived of a good blood supply, heat exhaustion can result. Symptoms often include heat cramps, a general weakness, moist skin (as opposed to dry skin in the case of heat stroke) and paleness of the skin (as opposed to redness in the case of heat stroke). The skin usually doesn't feel "hot" to the touch, and the body temperature is near normal.

Heat exhaustion can be serious. After administration of first aid, the victim should see a doctor. Like other problems related to heat, sun and lack of water which may occur in the desert, heat exhaustion can usually be prevented by sufficient intake of water and salts and avoidance of stressful situations in hot weather.

Heat Stroke

While all heat-related problems can be serious, **heat stroke should be considered a life-or-death crisis.** Accurate diagnosis of the problem followed by immediate first-aid are imperative. In heat stroke, the internal body temperature may soar to 105°F or more as the body's primary method of maintaining cooling (sweating) has broken down and is no longer functioning. A victim typically has a reddish face, and the skin feels "hot" to the touch. The skin will usually feel dry. Unconsciousness may occur quickly and other symptoms appear. It is imperative to get the services of a physician as soon as possible. Avoid stress, administer plenty of liquids and salts (unless the victim is unconscious), and avoid direct exposure to the sun's rays. (Note that heat stroke may also be referred to as "sun stroke.")

Never treat any heat-related problem casually. All require prudent and prompt first aid. Fortunately, most of these problems can be avoided. In most cases, it is the individual who fails to prepare who gets into trouble.

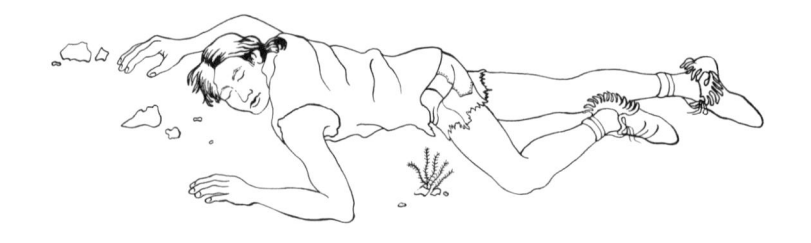

Prevention of Problems
Related to Heat, Sun, Lack of Water

As you gain experience in peacefully coexisting with the desert, you'll no doubt add to the following fund of information. But the techniques described below will help you guard against the development of an emergency or--if that fails--to avert disaster.

1. **Avoid stressful activities during the hot season:** In the Southwestern desert areas of the United States, the "hot" season usually extends from about April into October, but there is considerable variation so that very hot weather may sometimes occur at unexpected times of the year. (Personally, we consider a 90° F day as "hot," but this is an arbitrary figure.) Avoid heading out during hot weather. Death Valley once had 126 days in a row when the thermometer reached at least 100° F. Wait until fall, winter or spring if you plan an extensive outing on foot. Remember, however, that problems related to heat, sun and lack of water may occur in temperatures below 90° F although the risk increases as the thermometer climbs.

2. **Limit activities to cooler times of the day:** While our advice is not to head out on foot during hot weather, we confess to doing it on a regular basis anyway. When it is hot, however, we take shorter hikes, heading out before sunrise and trying to be back by noon. Besides being the coolest time of the day, many species of wildlife are still active, and flowers may be open which will close later in the day. The extra hour will also give you more daylight should an emergency require a longer return trip than anticipated.

3. **Keep your pace slow, and carry a light load:** Too many people who head out on foot or horseback seem more concerned with seeing how much ground they can cover in a day than with what they see and do. Unfortunately, they are missing a great deal. Rest often, take in the scenery, and never force yourself. If you're leading a group, maintain an even pace that is comfortable for everyone.

We have seen people carrying more gear than they could possibly need. Keep a checklist of items for both your daypack and backpack. Carefully weigh each item, and make notes of what you didn't use on each trip. Using this method, you may be able to pare down the list by eliminating items that you'd require only on special trips. We have even heard of people who have taken "lightness" one step farther by cutting down the handles on their toothbrushes and trimming the edges off maps. For most of us, there probably has to be a practical limit at some point. Nonetheless, keep in mind that the more weight you carry, the greater the stress and heat production. This increases the likelihood of problems.

4. **Wear suitable clothing:** In several of the case histories that we researched in the process of compiling this book, the victims had taken their shirts off because it was hot. This exposed their skin directly to the sun's rays so that their bodies soon absorbed more heat than their cooling mechanisms could handle. In Chapter 5 we have gone into the subject of desert clothing in more detail. Resist the impulse to acquire a tan while hiking or backpacking in the desert.

5. **Smoking will hasten dehydration:** If you find yourself in a survival situation, don't smoke; it will hasten dehydration and reduce endurance. While it seems a small matter, it could be enough to tip the scales in or out of your favor.

6. **Keep your mouth closed; avoid unnecessary conversation:** While talking will not greatly increase the amount of water lost from the body under normal circumstances, in a real survival situation where dehydration might be a problem, anything you can do (or not do) which will increase your chances should be considered. Avoid breathing through your mouth, avoid shouting and any unnecessary conversation.

7. **Avoid eating if water is not available:** Eating foods will only hasten dehydration if water is not available. Wait until you have plenty of water, then eat.

8. **Don't consume alcoholic beverages:** Alcohol will hasten dehydration. If you get out of an emergency situation, you can celebrate when you get home--being sure to have rehydrated first.

9. **Don't head out into the desert if you are sick:** Not only might sickness reduce endurance, cause dizziness and other problems, but it could also be accompanied by fever. Should you then become dehydrated, get into a stressful situation where considerable heat is produced or be overexposed to the sun, these factors combined with the already-elevated body temperature could be fatal. Vomiting or diarrhea can also result in a considerable loss of fluids from the body and hasten dehydration.

Never leave a sick or injured person alone unless it is absolutely imperative that you go for help. Be sure to fix the victim's location on a map that you carry with you, and mark the victim's physical location by placing jackets, etc., in bushes and trees. Be sure that the victim will be in the shade all day long, regardless of the sun's changing positions. If possible, situate the victim along a well-used trail so there will be no chance that a rescue party will miss him.

10. **Never head out on impulse:** Many victims of the desert perish because they took off (either on foot, horseback or in a vehicle) on the spur of the moment with few or no supplies, little or no water. Always be well organized, and plan the entire trip. Know when, where, why, with what and with whom you will be traveling.

Sucking on a pebble or chewing gum may relieve the sensation of a dry mouth, but it in no way adds to or stretches the water supply in your body. So it's up to you to decide, though there's an outside chance that it might dull your sensation of thirst and thus cause you to overlook one of the warnings that you're dehydrating.

CHAPTER 5 - CLOTHING FOR HOT WEATHER

If you're new to desert hiking and backpacking, the recommended clothing may come as a surprise to you. We did our share of hiking in shorts and sleeveless shirts, but we soon exchanged them for the more comfortable (and much wiser) long-sleeved shirts and long pants. In general, clothing should be light colored (white is the coolest because of its greater reflective qualities). **The most important function of your clothing should be to protect your skin from direct exposure to the sun.** Its secondary function should be to protect against cactus spines; thorns and twigs of mesquite, catclaw and other desert vegetation; and to keep insects from nibbling on you. One bare-legged hike through tall, dry grass will leave you itching and scratching and wishing you'd worn long pants. Also, heavy denim pants might help to deflect the bite of a rattlesnake. Following are short descriptions of those items we have found most useful.

1. **Hat:** Your hat should have a wide brim that goes all the way around to protect your eyes in front and your neck in back. Select a light-colored hat, and keep in mind that straw or similar types are the most comfortable in hot weather. Be sure to get one with ventilation holes. We have added a chin strap to ours for use when the wind picks up.

2. **Neck protection:** If your hat doesn't provide complete protection for the back of your neck, attach a piece of cloth onto the back of the hat. A large bandana is not only great for protecting your neck but has many other handy applications (sweat band, bandage, signal, wash rag, handkerchief, mask to tie over nose during dust storm, etc.).

3. **Long-sleeved shirt:** Not only will the long sleeves provide protection from the sun, but they will also help to protect you from scratches and insect bites. Be sure that the material is not so thin that the sun's rays will burn you right through the shirt. Cotton is probably the most suitable fabric as many of the synthetics have proven too hot to be practical.

4. **Levis or jeans:** Shorts and cut-offs are generally unsatisfactory if the weather is going to be hot and you are going to be out for any length of time. Remember--this is a matter of practicality as opposed to fashion or fad. A badly sunburned person is much more likely to find himself in a serious predicament on the desert than one who isn't.

5. **Boots:** Boots should be durable, fit well (very important because a poorly fitted pair can quickly ruin an outing), provide support, provide insulation from the hot desert ground and provide good traction. Don't cut corners by buying a cheap pair; get boots that will do the job and last. Soles should be thick and strong enough to resist penetration by such things as amole (a form of agave that grows low to the ground) and cholla stems.

CHAPTER 6 - COLD

Most desert areas are cooler and more inviting in the winter. In winter there is less likelihood of problems related to heat, sun and lack of water--but keep in mind that if you're careless, these problems can confront you at any time of year. Plan to carry water and salts just as consistently in winter as in summer. "Hot" weather (over 90° F) can also occur in winter. Even if the thermometer doesn't register this high in the shade, the overall effect can be the same if you have your shirt off to get some sun.

We know of an experienced desert hiker who has gotten into trouble twice on the desert. He heads out even in summer and has done so for years. One time he got into trouble in February and the other time in October--both times because he prepared more casually than he would have in summer. His experience carried him through both incidents, but he was luckier than a novice might have been.

Cold can be a serious problem in the desert, partly because people don't associate the desert with cold. It is not unusual to have warm, sunny days in the 70's and have the top of your sleeping bag covered with frost in the morning.

Hypothermia is not a common desert mishap, but it can and does occur. Every desert traveler should be thoroughly familiar with hypothermia, its symptoms and the necessary first-aid treatment. Remember, too, that most desert areas are bordered by mountain ranges which, sooner or later, most desert dwellers will visit. Here, at the higher and cooler altitudes, hypothermia is a very real possibility.

Hypothermia is a cooling of the body's inner temperature, from which death can result. Usually it is a result of several factors working in combination on your body, although occasionally just a single factor, such as cold temperatures, can bring it on. Cold temperatures (not necessarily below freezing), wet skin, wet clothing, wind and fatigue are usually contributing factors. When all of these factors are at work, you will probably be faced with the problem of hypothermia.

One of the major problems of hypothermia is that it can sneak up on the unsuspecting or poorly prepared individual. As the symptoms begin to mount up, the problem has progressed, and the individual may no longer be able to recognize his affliction or do anything about it. As the amount of cooling increases, it may be impossible for even the most knowledgeable individual to take steps to reverse it. Many victims of hypothermia (which occurs every year in areas such as Washington and Oregon where humidity and rainfall are great) have been found dead along the trail with sleeping bags, tents, parkas, stoves, food, etc., still in their packs. They simply sat down and were no longer able to function normally. As the cooling of their bodies increased, they drowsed and died. Had many of these victims been fully aware of what hypothermia is, they could easily have prevented it. At worst, their return home would have been delayed only a day or two.

Fortunately, **hypothermia can usually be prevented by keeping warm, staying dry, avoiding fatigue and staying out of the wind.**

The symptoms of hypothermia are often easier to spot in someone other than yourself. Immediate action is necessary to reverse the cooling trend, although you will learn in your first-aid course that there are some situations where rewarming must not be done too quickly. A numbness of the skin should warn you that you are cold and that at least part of your body is being cooled faster than body heat production can compensate for it. Symptoms will vary from one individual to the next and depending on the severity of the hypothermia, but staggering and careless walking could be symptoms. Watch other people in your group. Are they slipping, pausing or stumbling when they shouldn't be? Is your own muscular and mental coordination less than it should be? If conditions are cool, if there is a wind, if anyone is wet or tired, hypothermia could be occurring. Victims often slow down considerably. Their thinking may become muddled and slow. They may be unable to make simple decisions that would have been easy a short time earlier. Shivering is clearly an indication that your body is cooling. Your muscles attempt to compensate by shivering to generate more body heat. This mechanism often works if the heat loss from the body can be stopped. But if the heat loss continues, the shivering will increase the use of stored energy resources and create further problems.

What should you do if you suspect that you or someone else is suffering from hypothermia? Better yet, what do you do to prevent it in the first place? You will want to use as many of the following courses of action as necessary. Build a fire if dry wood is available. Get a good big one going, and once you are getting dried out and warmed up, stay with the fire until you are completely rested and certain that you can continue without being vulnerable to the same problems. You can also get warm by removing outer clothing and crawling into a warm sleeping bag. (Clothing restricts the escape of body heat into the sleeping bag so that the less clothing is between you and the insulated air in the sleeping bag, the more quickly your body heat will warm the interior of the bag.) In some cases, it may be necessary to crawl into a sleeping bag with a victim as his body may not be producing enough heat to prevent hypothermia from continuing. When the situation is this severe, it warrants your stripping down so that your body heat will escape into the bag to provide the necessary warmth that the victim's own system cannot provide. Drink warm liquids (coffee, tea, soup, water), and waste no time. If you can't find a natural shelter, build one; but expend as little energy as possible in the construction. Set up your tent, and retreat from direct exposure to the elements. Whenever you are hiking, nibble on high-energy foods which will reinforce your body's heat production capabilities. If you have had a problem and are once again warm and out of danger, consider spending the night in comfort and taking off the next day after full recovery and a good night's rest.

Keeping dry is one of the most important factors in preventing hypothermia. Wet clothing can contribute to a greatly increased loss of heat from the body. If you get wet and it is cool out, change clothing--fast. Build a fire, and be thoroughly warmed and dry before continuing.

A poncho and other rain gear are sometimes very helpful in preventing the body from getting entirely soaked. Some of the newer designs and materials "breathe," which enables sweat to escape and helps to keep you much drier and more comfortable than some of the materials used in the past. We always use the "layer" method, wearing several shirts and/or sweaters at one time. If the temperature changes, it's a simple matter to peel off what you don't need or to add more layers as needed. Be sure that the sizes are graduated so that one article fits over the other. Try to have at least one layer of wool clothing because wool will help to provide at least some insulation even if it becomes wet. Long underwear will provide you with another good layer of insulation. Also consider gloves and even a down vest and parka for some very cold desert areas in winter.

As you travel in cool weather, take frequent rest stops. Use your portable stove to heat up a cup of soup or coffee. Snack on energy foods, and always carry at least one extra day's food with you in case an emergency detains you.

Wind will greatly accelerate the heat loss from exposed skin. For this reason, always carry a wind parka in winter, just in case a wind rises. If you think that there is a probability of losing too much heat to the environment, find a sheltered spot to wait it out until the wind dies down. A good tent helps to increase the enjoyment of most wintertime desert outings, and it will help prevent hypothermia if you take the time to set it up and use it when needed.

Wind-chill factor is a number formulated on the basis of potential heat loss from the human body under various combinations of temperature and wind. We have included a simplified wind-chill chart simply to show you how to read one and to emphasize the important role of wind in heat loss. More complete charts are available from many sources. To read, simply determine what the actual air temperature is by using a thermometer. Find this temperature on the chart, then read across to the column showing what you estimate the wind speed to be. For example, if the thermometer reading is 20° F and the wind is blowing at 10 miles per hour, the wind-chill factor is 4--meaning the equivalent of 4° F. If the air temperature was 20° F and the wind was blowing at 20 miles per hour, the wind-chill factor would be -10. As you can see in these examples, an increase in wind velocity of only 10 miles per hour can result in a heat loss from the body the same as if the thermometer had dropped by 14 degrees. Or, with the wind blowing at 20 miles per hour and thermometer at 20° F, the potential heat loss is the same as if the thermometer were at 10 degrees below zero with no wind blowing.

Simplified Wind-Chill Chart

Actual Air Temperature ° F	Wind Speed in MPH			
	10	20	30	40
40	28	18	13	10
20	4	-10	-18	-21
0	-21	-39	-48	-53

Clothing for Cold Weather

We have already discussed clothing to some degree in our discussion of hypothermia. Cold weather gear for desert use in the winter is usually similar to that for cold areas elsewhere, although desert conditions usually are not as severe as mountain conditions, and you can usually get by with lighter clothing. Instead of a heavy parka, for example, a down sweater would keep you comfortable in most desert/winter conditions. Be sure to have a tough outer shell for down jackets and sweaters to prevent tearing them on vegetation.

The Great Basin and other "high" desert areas are usually much colder than the lower desert areas. **Always be prepared for colder and more severe conditions than you expect to occur,** then you should have no problems. Even very warm days can be spaced by bitter, cold nights.

Emergency blankets are metallized plastic sheets which are very light and fold into a small package. We usually carry two each in addition to our full range of winter gear. We consider them to be supplementary emergency devices, and they should never be thought of as substitutes for sleeping bags, shelters, etc.

In choosing a sleeping bag, check its temperature rating. These ratings are only approximate because the same bag will provide ample warmth at these temperatures for one individual while another might be very cold at the same temperatures. Don't be tempted to buy a cheap bag. It will not provide you with satisfactory service and is one short cut you can't afford.

Fires

If we plan on cooking, we carry a small stove that uses white gas. (There are also a number of small stoves that make use of disposable containers, but it is our opinion that too many of the empty containers get strewn over the landscape.) These stoves only weigh a couple of pounds and produce a considerable amount of heat. Pans and utensils keep fairly clean using this type of stove--certainly much cleaner than if you did your cooking over a fire. Be careful in the selection of a metal fuel container. We once lost most of our food when a faulty container leaked into the pack. Use a fuel container that can't be confused with water or food containers--and label it. We carry our stove and utensils in nylon stuff bags which keep soot and grease from getting loose in our packs.

If you need to build a fire, remember to use only downed wood. Green wood is more difficult to obtain, difficult to ignite, burns very poorly and may be illegal to cut in many areas (though in a real survival situation you would be delighted to see the arresting officer). Ironwood and mesquite are very hard woods and will burn relatively hot and slowly. **In an emergency or survival situation, if it appears that you will have to spend an unexpected night out, be sure to start gathering wood while it is still light.** If you do your collecting in daylight, you'll be less likely to close your hand over a scorpion or mistake a rattler for a stick of wood. Gathering wood after dark may be all but impossible if you are impeded by cliffs and other potential hazards. The large amount of wood required to keep even a small fire going all night out in the open would surprise you. You may have to resort to cholla stems and similar types of vegetation, and it will take you considerable time to amass enough to get you through the night.

Many people have never learned the skill (and it is a skill) of firemaking--especially under conditions where you cannot afford to waste time, matches or fuel. Practice in a campground or in your yard. The secret is to have plenty of dry kindling and all of your wood stacked and ready to go. Some people manage to get a small fire going, then race around trying to round up more wood before the flame goes out.

Hypothermia requires more study than we can briefly cover here. We strongly recommend that you pursue the topic further if you plan on heading into any cold environment. Be thoroughly familiar with the most recent first-aid treatments.

CHAPTER 7 - RAIN, FLASH FLOODS, LIGHTNING

A total annual rainfall in most desert areas of the American Southwest is usually around ten inches. This varies from year to year and from one location to another. Some desert areas in places such as Chile and Australia may receive no rainfall whatsoever for several years running.

In the United States and Mexico, most desert rains fall during summer. In winter there may be occasional slow drizzles that last for many hours, but during the summer monsoon season, rains usually begin and end suddenly and can pour down an inch or more in less than an hour.

In Arizona and other areas of the Southwestern desert, the monsoon season typically runs from sometime in June through September. During these months, mornings are often clear. Clouds begin to build up early in the afternoon, and--by about 4:00 p.m.--will let loose with torrential rains, lightning and thunder. Summer storms are more like heavy squalls which sweep across the desert floor, drenching everything in their paths and leaving adjacent landscapes bone dry.

Always anticipate heavy rains and lightning during the monsoon season. Watch for cloud build-up and storm movement on the horizons. Summer storms often move in rapidly and give you little opportunity to out-maneuver them if you're any distance from shelter. (**Never shelter in a wash** for reasons we'll explain later.)

Be sure that all your gear has rain protection. Special water-repellent covers can be purchased which slip over your backpack and sleeping bag. Also consider wrapping your camera and sleeping bag in separate plastic bags to provide additional protection.

Some experienced desert hikers carry rain gear in summer; others prefer to be soaked by the rain rather than by the sweat that usually accumulates beneath rain gear in hot weather. We would probably carry a minimum of rain gear unless we planned to be out over night or for an extended period of time. Even then, we would probably be carrying a small tent and rainfly which would preclude the need for rain gear. So the question of rain gear in summer is probably up to the individual.

Flash floods are a common desert hazard. They occur very rapidly after rain starts to fall, and it is not unusual for them to rip through areas where there has been no rain at all. A torrential rain a few miles upstream may dump enough water to turn a dry wash a few miles away into a raging flood. **All desert hikers, backpackers and drivers should be aware of the very real danger associated with flash floods.**

As the rain starts to fall during the monsoon season, the skies darken, thunder rumbles, and the wind begins to gust. A few big drops splat into the dust, usually followed by a deluge of rain and (quite frequently) hail. Parched desert soils are not primed to absorb the moisture. Tiny rivulets soon join others and empty into drainages. As the water continues to collect, these drainages race into the larger washes. Little creeks a foot or two wide merge and merge again to form a snarling wall of water, sweeping mud and debris in its path down a wash that may have been bone dry ten minutes before.

Water in these washes may be many feet deep, and even shallow washes may conceal deep channels. **Never attempt to cross a running wash unless you** are positive that the entire crossing is shallow with solid footing underneath. Even then, beware of the wall of water that may come sweeping down the wash. Vehicles easily stall in running dips and washes, and traction is often very poor. Heed all "dip" signs located along highways, and watch carefully for running water crossing even paved highways. It is always wise to wait for the water to recede or to take an alternative route. Even if nothing happens to you personally, a vehicle filled with sand and water is usually not worth much.

It is not uncommon for cars or trucks to turn up many miles downstream from the crossing where they were originally engulfed. Auto bodies, human bodies and personal possessions are often strung out over considerable distances. Some of them are never found.

Should you become stuck in a running wash, the Arizona Department of Public Safety suggests that you abandon the vehicle and move to higher ground. Their reason for suggesting abandonment is the imminent danger of being engulfed by a wall of water. But there have also been instances where people drowned while attempting to abandon a stranded vehicle. If the water is deep and the current fast, staying put would be your only choice. We can hardly recommend a specific course of action to resolve a specific emergency; we can list as many alternatives as possible and hope that through knowledge and luck you will choose the right one. In the instance of a flooded wash, we would rather wait for an hour or so until the waters subside before attempting a crossing.

Should you abandon the vehicle and reach safety, walk down the road far enough to warn oncoming motorists of the potential danger.

Hikers shouldn't cross running washes unless a dire emergency forces them to do so. Wait, sit down, find some shelter and enjoy the scenery--or take an alternate route. Even a strong swimmer would be no match for the currents in many desert washes; and flood waters usually drop almost as quickly as they rise.

Lightning is a potential hazard to the desert traveler. In fact, lightning is a greater threat to your life than rattlesnakes. It is common during the monsoon season and occasionally occurs at other times of the year. **Remember that a direct hit by lightning is not necessary for a strike to be fatal. It has a tendency to spread out on the surface after it hits, and a strike within a hundred feet could be fatal.**

Avoid prominence. Stay away from mountaintops, ridges and open areas where you are the highest point. Do not shelter beneath trees. Move quickly to be in a safe location before an electrical storm hits if at all possible. A vehicle is generally considered fairly safe because of its rubber tires. If you can't get to shelter, crouch atop a dry sleeping bag or small pile of rocks. This may provide some protection if you are close to a strike. Get away from anything metal, and place metal pack frames and similar objects at a good distance from yourself. Even objects which are not prominent may be hit from time to time, but at least try to have the odds on your side. Caves may be dangerous in some cases as current may pass through you as it passes from the top to the bottom of the cave.

Avoid being caught out in an electrical storm. **Consider shorter hikes in summertime, and limit these to mornings when the likelihood of storms is less than later in the day.** Check weather forecasts. Be familiar with the first-aid treatment for a person struck by lightning. Many supposedly "dead" people can be revived by immediate first aid.

CHAPTER 8 - WIND, SAND STORMS, DUST STORMS, DUST DEVILS

Many desert areas in the Southwest are subject to occasional gusty winds. These can cause serious problems for the driver and occasionally for the foot traveler as well. Not only can winds overturn vehicles, but they may also obscure visibility with dust and debris. Hot desert winds can increase the rate of dehydration. For a person already in trouble on the hot desert, a cool breeze could bring welcome relief, but a hot wind could compound his problems.

Drivers of high vehicles such as campers, trailers and vans should be particularly wary of high desert winds. These are often strongest where a highway crosses a wash or canyon where the wind is funneling perpendicularly across the highway. Roads may be closed to high vehicles for short periods of time. Heed all warning signs.

Sand and blowing dust probably present the most serious hazard to the driver. In certain areas such as the sand dunes near Yuma, Arizona, and in many agricultural areas where the land has been plowed, these dust storms are common occurrences. They can sometimes be spotted far in the distance while at other times they rise up out of the ground in front of you.

If the dust storm appears to be a brief squall, wait for it to pass. If it is more extensive and widespread, it would probably be best to turn around, find a spot for the night, and continue the next day.

Listen to your car radio for local weather warnings. These will usually mention dust storms in those areas where they are occurring. Arizona has instituted a "Weather Alert System" along certain vulnerable stretches of its interstate highways. Large signs which usually display directional information can be changed by radio control to issue warnings of storms which have been reported by officers in the field. A message is relayed from Phoenix which changes the appropriate signs. These signs were installed after some particularly large and gruesome accidents had occurred as a result of poor visibility during dust storms.

Many accidents which occur in sand and dust storms are rear-end collisions. These result when drivers stop or slow down without leaving the highway. Oncoming motorists are unaware of an obstacle until too late, and multiple-vehicle crashes are not uncommon. (The reduction of visibility can occur in just a second or two, so other drivers may have little or no warning.)

Besides danger to life and limb, consider damage to machinery from sand and dust. After traveling through a sand storm, it is a good idea to wash the vehicle, being sure to clean the underside. Change oil, oil filter and air filter. Sand may have sifted between the brake linings and drums, causing the brakes to grab unevenly. The windshield may become completely pitted or "frosted," making visibility impossible. Sometimes the side of the vehicle which bears the brunt of a storm will be sandblasted down to bare metal.

If you can't avoid a dust or sand storm, slow down (but not to the point where you might be hit from the rear), and turn on your lights. If visibility is poor, it is usually better to pull completely off the road and park. If you do this, be sure to turn your vehicle lights off so that upcoming traffic will not mistake your taillights for those of a vehicle still on the highway. **Never stop on the highway;** pull entirely off to the side.

We have seldom encountered problems with sand or dust while on foot, but when confronted with a storm of that type, the hiker should take the matter seriously. Try to find shelter such as the lee side of a large boulder or cliff (being sure there is no danger of falling rock). If you are stuck out in the open, lie down with the wind at your back. Protect your face with a bandana or an extra shirt. Visibility during a dust or sand storm is distorted so that landmarks are hard to distinguish and distances hard to judge. Consequently, you're probably wise to wait it out and proceed under more normal conditions.

"Dust devils" are desert whirlwinds that resemble tornados in appearance. On a hot summer afternoon, you may be able to see several at a single time in some desert valleys. They are not true tornados, nor are they as powerful. In general, they fit more into the category of inconvenience as they loft leaves, trash and dust to a height of several hundred feet. A very strong one could knock you down if you weren't ready for it, but we've never heard of that actually happening to anyone. Seek shelter or protect your face for the few seconds it takes for one to pass by.

CHAPTER 9 - MISCELLANEOUS HAZARDS

Diseases

Most of the usual diseases of man can be found in desert communities, but the desert itself is a relatively healthy environment. Normal precautions should be taken, and you should make every effort to avoid contact with native animals. Many people insist on killing coyotes and other predators, but the overall results are increased rodent populations and increased risk of contracting diseases from their parasites.

Rabies occasionally occurs in desert mammals as it does in mammals elsewhere. County and state health departments have been notorious for irrational "control" programs such as skunk irradication. The overall effect is negative for man as it is for skunks, and most of these programs should be abandoned. Mammal populations will undergo a natural fluctuation from time to time; but predators, disease, space and availability of food and water will act as natural controls. Simply **leave all desert mammals alone.** Avoid the temptation to pick up "lost" young (most are no more "lost" than you are), and by all means, never contact an obviously sick animal. Bats and bat caves in particular should be avoided.

Bubonic plague is endemic in many desert areas. Avoid camping where rodent holes are common, and don't trap or pick up potential carriers. The disease is transferred to man by certain species of fleas which leave the body of the host and then bite man. There has been a slight increase in reported cases of plague in recent years. Reductions in the natural predators of potential carriers such as ground squirrels is one definite factor in this increase, as are carelessness and lack of knowledge.

Bites of Animals

In Chapter 21, we cover certain aspects of bites by venomous animals. **Most animal bites occur when people handle animals which should have been left alone.** After providing necessary first aid, contact a physician. Avoid having anyone else bitten (or the victim being bitten again) in an attempt to capture the offending animal.

Cacti

Cacti can be a problem in many desert areas, as can certain species of agaves and some yuccas. The usual culprits are cholla stems. There are a number of species of cholla in the desert, but two species in particular have very stout, sharp spines. These are the Jumping Cholla (Cylindropuntia fulgida) and the Teddy Bear Cholla (C. bigelovii). Their stems often litter the ground and are hard to avoid in some areas. They can penetrate all but the stoutest of soles and can usually pass through the leather sides of boots. Be careful not to brush against cholla cacti because even the slightest contact is enough to detach stems which imbed in your skin. In the minds of some of us, "cholla" and "pain" are synonymous.

If a cholla stem is deeply imbedded, it may be necessary for a doctor to remove it. It can usually be removed by slipping a pocket comb between the stem and the skin and flipping it away from the body, being careful not to flip onto a bystander or another portion of your anatomy. Pliers may sometimes be helpful, and tweezers are handy for removing individual spines and patches of tiny spines known as "glochids." The latter are found on prickly pear pads and cholla stems. A magnifying glass may be necessary to track down all of these little spines.

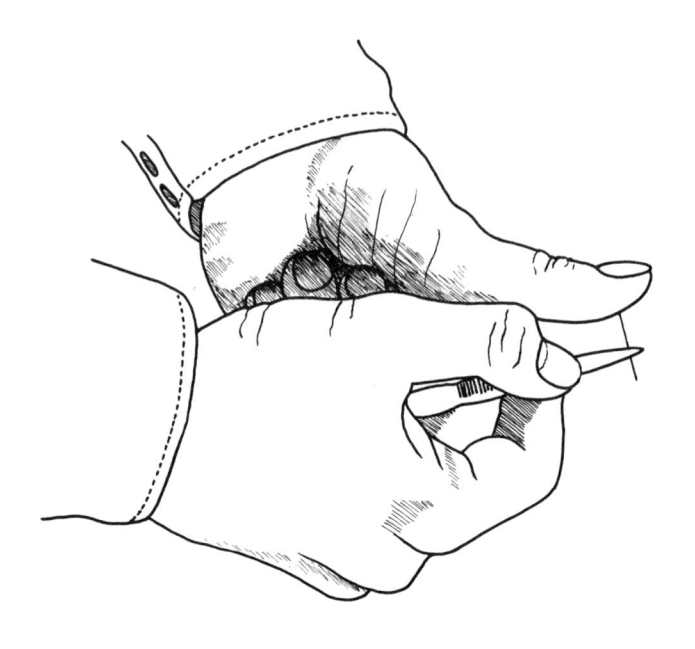

CHAPTER 10 - FOOD

In our opinion, food is seldom a factor in survival during hot weather, assuming that your salt intake is sufficient. Lack of food and nutrients is much more likely to cause problems in cold, winter weather. For this reason and others, we are de-emphasizing a favorite topic of most survival books--the gathering of native plants and animals for food. We have seen long lists of so-called edible foods including such impractical gourmet items as mountain lions and hawks.

There is no question that experienced individuals can go out and, given the proper time and location, gather a wide variety of wild, edible food in the desert. For a novice to go out and do this with any degree of success while in a real survival situation would be a questionable expenditure of energy. There is a great risk of finding little or nothing edible, squandering precious energy in the attempt, dehydrating, getting lost or hurt, etc. **The risks must be weighed against the probability of getting nutritious food and the real need for such food. Most desert survival cases are solved for better or worse within a short length of time--too short for food to be a major concern. The real desert killers are heat, lack of water and salts, accidents and ignorance rather than starvation.**

Most people can go without food for several days without suffering much more than hunger pangs. Assuming that you are in good physical condition and have plenty of protection, shade, water and salts, you should be able to get through four or five days before having to consider food a problem. Food would be a more immediate problem in cold weather, but the chances of finding nutritious food at this time of year are usually much reduced.

If you decide to attempt to gather wild foods in a survival situation, weigh the potential energy expenditure against the probability of getting certain foods and the energy they will provide. Spending half a night to catch a mouse, even if you were successful, would be of doubtful value if that mouse provided less energy than you used in the pursuit and capture. Also consider such factors as the palatability of the food, edibility, possible diseases the animal might carry or poisons the plant might contain, toxins, abundance and preparation needed. Being hungry or uncomfortable is one thing, but risking your life unless starvation is a real possibility is another.

Always drink water with your food. **Eating food without water can increase dehydration and result in other problems.**

Besides food for regular meals, carry plenty of snacks plus additional food for emergency use in case you are stuck out unexpectedly. We try to eat a big meal before heading out, and we like to stop at a restaurant on the way home for another big meal. In between, we usually eat rather lightly. We always prefer to take along foods which don't require cooking, thereby eliminating stoves, fuel containers, fuel, eating utensils, soap for cleaning and stuff bags to hold all these items. As a result, our loads are lighter, and there is more room in our packs.

Hard candies, tropical chocolate (higher melting temperature), jerky, cheese, crackers, salted nuts, dried fruits (raisins, apricots, pears, apples), fresh fruits, pretzels and the like make excellent trail fare. We try to combine sweet and salty foods. Cans and bottles should be left back at home or in your vehicle; once empty, they're a nuisance in your pack and a temptation to litter.

Wrap all food carefully, and place it in a safe location at night where skunks and rodents won't get into it. One mouse can make a fair mess of things, and an ambitious wood rat can imperil your entire larder. We know from what remained of the food in our backpacks one morning that there is at least one very obese mouse waddling around in the Grand Canyon.

Theoretically most desert mammals, birds, snakes and lizards (there is some question in our minds about toads) are probably edible. In addition, most kinds of birds' eggs and many species of insects could provide nutritious meals. We feel that most have not been tested, and there is always the possibility of parasites or diseases or getting sick from toxic parts of some "foods." In addition, most of these animals would require various degrees of preparation to make them edible, and there is always the danger of being bitten in the process of collecting them.

The list of possibly-edible desert plants or parts of plants is long; entire books have been devoted to the subject. Some of these books have not been written from firsthand experience, and we have come across a number of questionable statements in some. Again, many species of desert plants haven't been extensively tested to determine whether or not they are nutritious or whether or not they may contain toxic substances. Some so-called edible plants may also have some poisonous look-alikes. Of course, we have collected such things as prickly pear fruits (strictly a seasonal food) to make jelly, and the Papago Indians annually collect the fruits of saguaro and organ pipe cacti for making preserves and wine. We have also eaten a number of other plants from the desert, but they were ones with which we were thoroughly familiar.

In short, we believe that **food is seldom a factor in short-term desert survival in hot weather.** This is assuming that you have plenty of salts and water. While there are instances where a person in a survival situation could gather certain wild foods to provide nutrition, we suggest that you do so with caution and carefully consider the reliability of the author or the person recommending the plants you intend to consume. Always carry plenty of food plus an extra day's supply, then you probably needn't concern yourself with snaring a hawk or trapping a mountain lion.

CHAPTER 11 - SHELTER

In a survival situation, shelter is a vital requirement for protection from wind, sun, rain, animals [possibly], sand, cold and heat. The simpler the shelter you can find or construct, the less energy you'll expend and the less water your body will lose. Look first for natural shelters. Later, if you must stay in a particular area for some time, you can find another shelter or build something more elaborate.

Caves can provide an immediate shelter in some areas; however, they have some drawbacks. Wood rats, rattlesnakes and other animals often reside in them. Odors may be strong, cholla stems scattered about, and parasites of certain animals may be common. Steer clear of bat caves as bats and bat guano are likely carriers of disease. Some caves, especially in mountainous areas, may be unsafe during electrical storms. Cliffs and large boulders may provide good temporary protection, but first rule out the danger of falling rocks. Don't set up a tent or shelter in a desert wash unless no other shade or protection is available. The danger of flash floods is too great, especially in summer. If you're forced to shelter in a wash, be sure to abandon the site if a storm is brewing, either locally or upstream.

Emergency shelters can be contrived from materials such as branches or brush. Check your construction materials carefully to be sure they are free of scorpions and other invertebrates. Many people carry some rather basic items from which an emergency shelter can be constructed. Tube tents are made of polyethylene plastic and are tied at both ends. They are available at most outdoor stores and can provide fair protection from rain but are not much help in protecting from the sun unless other materials are incorporated into the design. Some people rely on a simple sheet of polyethylene plastic, though our experience has been that unadorned plastic is too flimsy to be serviceable for long. Another favorite is a specially-coated nylon fabric that provides a little more durability. Your emergency blankets can also be incorporated into many kinds of temporary shelters, but these, too, are easily torn. Be sure to carry plenty of cord for purposes of rigging your shelter, and we carry vis clamps (little metal loops with rubber balls) as a means of anchoring the cord to the fabric without having to punch holes in it. Any method of anchoring the cord to the fabric is preferable to punching holes that will rip out in a breeze. In a survival situation, use any materials at hand (maps, clothing, sleeping bag, etc.) to provide emergency shelter from the elements.

Once you are in an emergency shelter, **try to elevate yourself off the hot ground.** Sit atop your sleeping bag or on a pile of clothing or wood. Even getting a half foot or so above ground level could increase your chances.

When selecting the site for an emergency shelter, keep in mind the sun's changing position as the day wears on. Try to choose a site that already provides some shelter yet affords quick access to an open area for signaling search planes or rescue parties.

A good, durable tent is recommended for most desert backpacking. Select one that has a sewn-in floor and a rainfly. The floor is necessary to keep out unwanted visitors (scorpions, etc.); and the rainfly, besides its obvious function, will help to keep the inside of the tent a little cooler. Be sure to carry extra cord because tent stakes may not hold in loose desert soils. Tie the cord ends to heavy rocks, tree trunks, etc.

One type of tent which is really comfortable and practical for hot-weather camping has sides constructed of mosquito netting and has a rainfly for wet weather. Some tents have a built-in rainfly which simply rolls up to expose the netting. The netting permits breezes to circulate through the tent while keeping gnats and other pests out. If the night is hot and insects are a problem, this type of tent will pay for itself quickly in comfort alone.

When constructing an emergency shelter in winter, try to incorporate a fire into the design, but keep the fire outside the actual structure.

CHAPTER 12 - EQUIPMENT FOR FOOT TRAVEL

Good equipment and a knowledge of its use can enhance the enjoyment of any desert outing and provide you with good tools for use in an emergency. On the other hand, too much or poorly-selected equipment can be more detrimental than helpful. In many cases it has been the poorly-equipped, poorly-prepared individual who met with serious trouble in the desert. We once met an elderly man dressed in a dark blue suit and street shoes about ten miles down in the Grand Canyon. He hailed us cheerily, asked how far to the river and continued on. He had no water, no food, no hat. The elevation gain coming out of the Canyon is some 4,500 feet, and we've often wondered how he got back out.

We have covered an assortment of equipment in other sections of this book. Here we cover what we personally consider to be essential gear for daypacks and backpacks. A few optional items are also discussed.

Daypack and Emergency Kit

On all short desert outings, we carry daypacks with emergency kits. Included in the daypacks are items such as canteens of water, extra socks, extra boot laces, a sweater or down jacket, a windbreaker, camera and accessories, binoculars, insect repellent, toilet items, litter bag, guidebooks, maps, permits, sunglasses, food for meals, food for snacks and extra food for emergencies. These items are tailored to our individual needs, but the emergency kits go into each pack regardless of how short the trip may be or how lightly we intend to travel. Other items--such as canteens--always go with us, too; but they do not fit into the emergency kits. Each person should have his own complete emergency kit, and each person (children included) should carry his own emergency kit, his own water and other vital necessities.

Our emergency kits consist of the items listed below, contained in a plastic "case" with a zipper closure. It is water proof and brightly colored to insure against its being left behind accidentally. Make a checklist of all items in the kit, and always check them all off before heading out. Recently we have accumulated so many items that we have each added a second kit of emergency items. Our personal list of emergency kit contents is as follows:

Batteries (extra sets for flashlights)
Blanket (metallized, emergency)
Bulbs (spares for flashlights)
Can opener (small G.I.-type)
Candies (non-melting)
Comb (for removing cholla stems)
Compass (quality important)
Cord (nylon)
Dimes (for pay phones)
Elastic bandages
Fire-starting jelly (commercial compound in tube)
First-aid kit (complete enough to meet all common emergencies)
Flashlights (**two** small ones per person)
Knee brace

Knife (pocket-type with a variety of blades)
Insect repellent
Lighter (disposable, butane-type)
Lip salve
Match containers (two, waterproof)
Matches (paper)
Matches (waterproof)
Moleskin
Money ($20 bill)
Needles and thread
Notebook
Pencil
Prescriptions (extra supply in case of delay)
Razor blade (single edge)
Safety pins
Salt tablets
Shoelaces
Toilet paper
Tubing (small diameter, about three feet long, rubber or plastic)
Tweezers
Vis clamps
Whistle

Other items which we occasionally add to our kits:
Bags (large, plastic)
Candle
Fishing gear
Flint stick and steel
Keys (car; house)
Lens (magnifying)
Rubber bands
Saw (wire)
Tea bags
Thermometer (in a case)
Tube tent
Watch
Wire

Backpacking Equipment

The following is a list of items which we usually carry on a backpack into a desert area (though the cook stove and utensils are optional items with us). Note that for colder weather a warmer sleeping bag and additional clothing will be necessary.

    All items which we would carry on a daypack except for the daypack itself
Backpack (durable, comfortable)
Cook stove (small, white-gas type)
Cooking and eating utensils
Food (discussed elsewhere)
Ground mat (roll-up type for insulation and comfort)

Miscellaneous items (such as climbing gear, ground cloth, etc.)
Litter bag
Tent (sewn-in floor, good materials and construction, perhaps with mosquito-netting sides for hot weather use; rainfly; cord; stakes; poles)

Again, we have a checklist of items which we run through before heading out each time. Weigh all items, and keep notes on what you use and don't use. The list should be a flexible one which changes to meet new conditions and changing personal experience, likes and dislikes. Never give in to the temptation to just toss things together. You'll wind up with too many items, too much weight--and you may overlook some very important item.

CHAPTER 13 - MAPS

Some desert hikers and travelers seldom if ever use a map, preferring to strike out and take things as they come. This can be a lot of fun, but maps can certainly contribute a lot to the enjoyment and safety of an outing. Mines, geological features, cliffs, washes, springs, buildings, creeks and other items are marked on many maps. With a map, you can pinpoint your present location and select feasible routes to various destinations. Directions and the prudent use of a compass in conjunction with a good map can help eliminate getting lost. On some older maps, man-made features such as roads and trails may no longer be accurate, so try to have the most current information available. **Try to obtain all the maps and information you need for a particular trip before leaving home.** Ranger stations and similar sources of information may be closed on certain weekdays, weekends or at certain seasons of the year.

The United States Geological Survey maps (commonly referred to as topographic or "topo" maps) are usually the best and most detailed sources of information for many areas. Check the date on the topo map, and keep this date in mind when referring to man-made features. We have occasionally found errors on maps, but it hasn't been a problem as we never depend completely on maps for following a particular route. **We rely on maps as supplemental sources of information.**

The USGS publishes a master map for each state. This general map shows the various maps which cover particular areas within the state. Usually they are designated (and ordered) by a name and a date. Most maps represent either a 7½- or a 15-minute quadrangle. On a single trip, you may need several different maps for complete coverage of your route. Ask your map dealer or the USGS for the sheet which interprets the various symbols on the topo maps. Many outdoor stores and stores that carry surveying equipment sell a variety of topo maps for local areas. **Maps can also be ordered by mail from the United States Geological Survey, Distribution Section, Federal Center, Denver, Colorado 80225.**

Contour lines on topo maps represent changes in elevation. The "interval" (amount of elevation change from one contour line to the next) is marked on each map, with either 40 or 80 feet being the commonly-used increments. The closer the lines are together, the steeper the terrain; the farther they are apart, the more level the terrain. These maps are easy to learn to use. The best way to become familiar with them is to have someone show you how to use one in the field where you can compare the terrain with the details on the map.

Other government agencies also provide various maps of local areas. These are seldom as detailed as topo maps, but they may be more current. We have occasionally had to consult several maps at one time. Service station maps are seldom of much value for travel into remote areas. Verbal or written inquiry may be necessary for information on some isolated spots.

CHAPTER 14 - LOST IN THE DESERT

Introduction

The word "lost" has many different meanings, ranging from being temporarily disoriented or being temporarily unable to find the route to having no idea of exactly where you are or which way to proceed. We all lose a trail from time to time--stretches may have been washed out by rains, the track may fade out over rocks, it may be obscured by brush, the trail sign may be adorning a bedroom wall in town or lying face down a quarter mile from the trail, or cattle may have tracked up the area.

Very few trails are clear for their entire route, and we often have to do some reconnoitering before we can continue. Once in a great while, we find that we must abandon a planned route and return the way we came. This most often happens in areas where we are bushwhacking or on old, abandoned trails. Impassable cliffs, dense vegetation such as catclaw thickets and similar obstacles can result in your not being able to follow a planned route. Also, landmarks in the flat desert may be visible for long distances, but those in rugged desert canyons may elude you entirely. Your life may depend on your turning back if you can't spot the trail or a vital landmark.

If you are lost to the extent that you can't find the route in a few minutes, sit down and relax and think things over. Never give in to panic. If you are fatigued, dehydrated, lacking salt and nutrients, or if you are hot, your ability to reason may be impaired. Drink some water, seek some shade, get some salt into your body. What may seem like an easy decision to make as you sit reading this book at home may be very difficult to make under actual field conditions. Where was the last spot where you were definitely on the trail or recognized a specific landmark? Can you hear the sounds of traffic or machinery which might help you get directions? What do you anticipate ahead? What is the status of your water supply? Of your food supply? Are you in good condition and able to proceed easily? What are the conditions of other members of your party? Is anyone injured? You must consider the weakest member of the party first. Will people be looking for you, or did you change plans without notifying anyone? Did you leave a note with someone? What is the thermometer reading? Is the wind blowing? What time of day is it? You must now ask yourself all these questions and others, then evaluate the situation and take necessary steps to get out of your predicament. **The more knowledge and experience you have and the better equipped you are, the greater the likelihood that the action you choose to take will get you and the other members of your group home safely.**

Being lost or having to spend an unexpected night or two out can be a fatal experience for one person and an invigorating challenge for another. **Never panic if you feel lost.** There have been instances where people have dropped everything they had--including canteens and sleeping bags--and just started to run. By the time they became exhausted, they were even more disoriented and less capable of handling the situation than if they had remained calm and stationary. Don't compound your problem by acting thoughtlessly or carelessly. Small children sometimes do better in "lost" situations than do many adults simply because they don't panic. (On the other hand, there are records of lost children who crossed paved highways and continued on into forest or desert on the other side.)

Don't become further lost. Obviously you got yourself in here, and certainly you can get yourself out. **Maintain a positive mental attitude at all times. Don't waste your mental abilities on negative thoughts of "what if . . .?" Concentrate only on the best and quickest way by which you WILL get out.**

Not Getting Lost in the Desert

If the route ahead is obscure, or if the day is well advanced and you aren't sure that you can complete the trip as planned, **consider retracing your steps.** For one thing, you should be familiar with the backtrail, having just passed over it; you KNOW what's in store by that route. The trail ahead might harbor some unfortunate surprises if you haven't the time or aren't equipped to cope with them. If nothing else, you are more fatigued at this point, and the temperature may have soared as the day wore on. Descending a mountain, especially if you're on a good trail, is usually quicker than ascending--but be warned that many mountaineering accidents occur on the descents. Don't be less cautious or pick up too much speed on the downhill stretch. Hiking in a place such as the Grand Canyon involves mountain climbing in reverse. The return trip is the uphill part and usually takes about twice as long as the hike down. This is a rule of thumb. Many people find the trip out of the Canyon impossible, and mules are sent down to haul them out (for a fee).

Before you start out, mark your entire planned route on a map. Circle landmarks (and make a point to spot these as you travel). Check with friends who may have been over the same route; they may be able to clue you in on difficult stretches and obscure places. Check compass readings, and try to visualize the entire trip.

If you are out with a group, have the leader signal the rest of the party as soon as the route becomes obscure or lost. This way, at least one person stays back on the trail so that it will be easy to find the route back if it becomes necessary to turn around. Be sure that everyone stays in voice contact. Each member should have a whistle and know the prearranged code system should it be necessary to use it. Once the leader has found the route, he can signal the rest of the group to continue.

Methods of finding directions in the desert are many, and we can't cover them in great detail here. The more methods in which you are proficient, the greater the probability that you will be able to put one of them to use in an emergency situation.

Landmarks are one of the best and easiest methods of determining locations. Formations such as mountains, canyons, washes, cliffs and pinnacles can all be identified on many topo maps.

A **compass** is a very useful tool for those who are proficient in its use, but it can be a hinderance for those who aren't. The user must take into consideration the magnetic declination for his locality. (The magnetic declination is usually marked on topo maps.)

The **sun** is fine for getting a very general bearing for assurance that you are heading in approximately the right direction. Remember that it rises in the east (approximately) and sets in the west (approximately).

The **North Star,** like the sun, is good for getting a general bearing if you must travel at night. It might be useful (for example) if you were crossing a wide valley and knew that a highway crossed from east to west. Find the "Big Dipper." Then find the two stars that form the part of the "cup" farthest from the "handle." Visualize the distance between these two stars, then look about five times this distance in a straight line running through these two stars and away from the open part of the cup. The star you see is the North Star. A line from you to that star points within one degree of True North.

It takes experience to learn how to **judge distances** in the desert. Mountain ranges which may appear to be only a few miles away from you can turn out to be 20 or 30 miles away. Distances across wide valleys are especially difficult to judge.

Solo hikes into the desert are not recommended, although this is not to say that many people don't do it. There are certain advantages to hiking alone, such as self reliance, ability to set your own pace and the fact that you usually spot more wildlife. Those rugged individuals who hike from the Canadian border to the Mexican border usually hike solo. The main problem with hiking alone is that even a relatively minor injury such as a sprained ankle can be serious without a companion to go for help.

If you are hiking from a highway or road, remember which way you must walk to return to your vehicle once you return to the road. If you hiked to the right and your vehicle was parked to the left of where you came out on the road, the consequences could be serious if the weather was hot or if you were dehydrated.

There are thousands of roads in the desert, ranging from parallel ruts to super highways. Many side roads and intersections in remote desert areas are not signed. **Keep a road log** as you travel. An innocent turn or two may be the introduction to a labyrinth of backroads. If you log your mileages and note down whether you turned left or right, you can extricate yourself with relative ease.

The Night Out

Let's say that for one reason or another you have to spend an unexpected night out. Maybe the route was longer than you anticipated and darkness set in or something else delayed you. If you are backpacking, there should be relatively few problems as you will probably be carrying enough gear, food and water for the extra time out. If you don't have a sleeping bag or water, you may be in for a bad night.

Day hikers often spend all their daylight hours trying to get out. As a result, they have to establish their emergency camp in darkness. If there is a full moon, there may not be a serious problem; but if it's dark and you don't have a good flashlight, setting up a decent camp can be difficult if not impossible. **Once you realize that you won't make it out until the next day, make an effort to find a suitable shelter, and gather plenty of firewood if the temperature is going to drop. The secret is to have at least half an hour before darkness to set up your camp.** If you are not already cold, consider waiting until darkness sets in before lighting your fire. This will help make your wood supply last longer. Searching for firewood in darkness is risky. There is danger of falling over cliffs or tangling with snakes, scorpions, cholla, etc.

When you're safe at home, darkness doesn't seem menacing, but out in a survival situation, many people find darkness a real foe--usually more imagined than real.

We have discussed night hiking elsewhere. Certainly there are many potential problems--the possibility of walking over a drop, plunging into a mine shaft, difficulty in recognizing landmarks and the possibility of encountering a snake or a skunk. We seldom hike at night, but if we do, we choose a well-marked route and make sure everyone has a flashlight.

Sleep is very important for normal functioning and especially important in survival situations. It is needed for maximum physical efficiency and clear thinking. Yet in a survival situation, it may be all but impossible to sleep. With two or more people in a group, one person can stay awake for a shift and keep an eye out for rescuers, keep a fire going, etc., while the others sleep. When you are stuck out alone, this is difficult to do. Try taking a nap in a shaded spot during the day, and at least **get as much rest as you can even if you can't sleep.**

Psychological Factors

Interviews with survivors of emergency situations have indicated that, **time after time, it's a person's mental attitude that makes the difference between surviving or perishing.** Those individuals who think they will survive often do while those with a negative attitude often don't make it. Certainly your state of mind makes a difference, at least in some cases.

People who have been stranded or injured or lost in remote areas (and even in areas near cities) have reported that a number of psychological factors often come into play. These are often things that you give little thought to until you are stuck out, but even experienced outdoorsmen report experiencing them from time to time. Many of these psychological factors can be expressed as various kinds of "fear." Remember that should you find yourself in a tight situation, experiencing these fears is a normal reaction. **Anticipate that you will experience fears, then be able to recognize them for what they are, and you will be more capable of dispelling them.**

We have already mentioned that **darkness** is one thing that many people find themselves very concerned about. A twig snapping in the darkness, not being able to find something, all help to trigger this fear of the dark. Many survivors have reported that **loneliness** was the most significant psychological problem they had to face. Some report a fear of **dying;** but this is a passing thought for most people. Many people in emergency situations are overly concerned about the danger of personal **injury or harm,** even though there is no real reason for concern. Others are very concerned about **animals.** The howl of a coyote nearby may terrorize one person and amuse another.

You will probably encounter some of these mental fears if you find yourself stranded over night. Remember that most of the fears reported existed only in the minds of the people questioned, not in reality. Once you realize this, you should be better able to cope with imaginary fears. You may not experience these fears, or you may find you have to contend with one or two of them. Anticipation and recognition are big steps towards overcoming fears. Be prepared, carry good equipment, be in good physical condition, get out often, set reasonable goals, and reduce the unknowns as much as possible. Keep a positive mental attitude at all times. Always think things through, and avoid impulsive actions motivated by imaginary fears. Fall back on your training and experience to get you out of a tight situation.

Group Survival

All factors which have been considered in individual survival also apply to group survival. Groups may have certain advantages in survival situations, but they also pose special problems. We don't like to head out with groups and very seldom do. We like to set our own pace and stop often along the route to examine this or that. The more people you have, the greater the probability that one will become sick or injured, the greater the chance of someone's having psychological problems, the more shelter will be required, the more water, more food, etc. If shortages occur, problems are compounded. A small water seep might be sufficient for one or two people in an emergency, but it may be of little value to a group of ten or more. Keeping the noise down, setting the pace and steering conversations away from politics and philosophy are special problems that accompany most group endeavors. It's a rare, good group that doesn't have at least one dodo, one hero, one Plato and three leaders (one acknowledged and two usurping). There is always someone who forgot something and has to borrow or who just got a new pair of boots that are crippling him.

Group outings are usually more successful if there is a strong, knowledgeable leader who can **resolve problems before they get out of hand. In survival situations, all members of the group must work as a cohesive unit.** Tasks can be assigned depending on individual skills, but all must keep in close contact and help other members of the party.

There are some advantages in traveling with a group. The more members on an outing, the more likely a note has been left and that a search party will be organized. Groups may be more easily spotted from a distance than a lone individual. There is also a better chance that someone will remember a crucial part of the route. The signaling system, with the leader calling out when the trail becomes lost, works best with groups.

Never send an ill person back to a trailhead alone unless there is absolutely no alternative. Such a person may pass out on a hot trail and quickly perish as his body rapidly absorbs great amounts of heat.

Rescue and Signals

Rescues usually take time. Party members must be called out, organized, and information verified. If the individual is reported overdue but didn't leave a message with someone (see Chapter 15), the rescue often takes longer. **A hiker who fails to leave a note and has no one to miss him may never be the object of a search.** Remember that most search and rescue personnel are volunteers. Usually they are not sent directly out unless it is known for certain that someone is injured or needs help or that a small child is involved. Most people reported overdue show up of their own accord before the search gets organized (or shortly after it gets started). Some thoughtless types forget to announce their return, causing a lot of unnecessary work and embarrassment. So take delays into consideration when writing your note, and don't forget to tell the holder of your note when you're safely back.

Most rescues take place within the first 24 hours, although there have been rescues that took place weeks after the victim became lost or otherwise unable to get back on his own. There are few people who survive alone for weeks in the hot desert, but it can be done.

After considering your immediate needs (water, shelter from the elements, food and such), **make every effort to increase your visibility so that searchers will spot you.** An aircraft may take only a few seconds to pass over the narrow canyon where you are. If you are not ready to signal when you first hear the craft, it is not likely that you will be spotted. Consider from what direction help is most likely to be coming. Will they be on foot, horseback, in a vehicle or a plane? Whatever signaling method you use, three of anything (gunshots, whistles, etc.) is considered to be a distress signal. The reply is two of anything. Consider the following as possible methods of attracting attention or signaling would-be rescuers.

1. **Lay conspicuous items [clothing, aluminum foil, tents and whatever you have] out on the ground.** Hang some of these items from trees where they can be spotted from a distance. Tie a bright shirt or some other material which contrasts with your surroundings onto a long stick such as an ocotillo stem, and be ready to wave it around should you see or hear potential help in the distance. Leave this stick upright so that it will be ready to use and so that it can be spotted from a distance in case you can't get to it in time to manipulate it.

2. **Use your emergency blanket as a giant reflector.** The ones we have are in a small packet but unfold into a sheet measuring 56 inches by 84 inches. Anchor it down with rocks, or tie it to a stick like a flag.

3. **Build a fire, and keep it going.** Have a small fire going, and have plenty of materials to toss on it should you spot help. Fires have the added advantage that they are often reported by people who do not realize that anyone is lost. If possible, use three fires in the form of a triangle. This is usually feasible only if there are several people in a group; one person would expend too much energy and lose too much sleep trying to keep them fueled. Be very careful to avoid the heat if there is any danger of dehydration due to high air temperatures, a water shortage, etc.

On clear days, dark smoke is usually more visible than light smoke. Materials available in your vehicle (seat cushions, motor oil, transmission fluid, tires, floor mats and insulation on wires) will usually produce a dark smoke. On the other hand, light-colored smoke may be more visible on dark, cloudy days. Green vegetation usually produces a light smoke when heaped on a fire.

If you use a fire for signaling, keep a small one going at all times (unless dehydration is a problem), and be ready to toss additional fuel on it if you think someone might see it. Keep in mind that if would-be rescuers spot a fire at night, they will probably try to check it out first thing in the morning. Be prepared to signal as strongly as possible if searchers should be headed your way come daylight.

4. **Try shouting and yelling,** but only occasionally or if you know that help is near. Never yell unnecessarily as this will result in additional dehydration and fatigue. "Help!" (or "Socorro!" in Mexico) is probably the best thing to yell. Shout at well-spaced intervals; then listen carefully for a response.

Whistles carry sound much farther than yelling and are less tiresome to use. Buy a good, loud one that will last for years. Even two rocks hit together sharply will make a sound that carries a fair distance.

5. **Hunters can fire shots in sequences of three.** A slight pause between the first shot and the following two will enable searchers to zero in on the direction of the sound. Shots may not be noticed during the daytime as they have become a rather ordinary sound in the desert, but they may be quite effective at night.

6. **A signal mirror** is an excellent device for attracting attention if rescuers are coming from the right direction on a clear day. Most of these

mirrors are made of metal and come in a case to prevent scratching. We carry one at all times and have found that the signal can be seen across long distances if used properly. If nothing else is available, consider a piece of foil, a tin can lid, mirrors from your vehicle or even shiny hubcaps. Signal mirrors are often very effective in signaling aircraft.

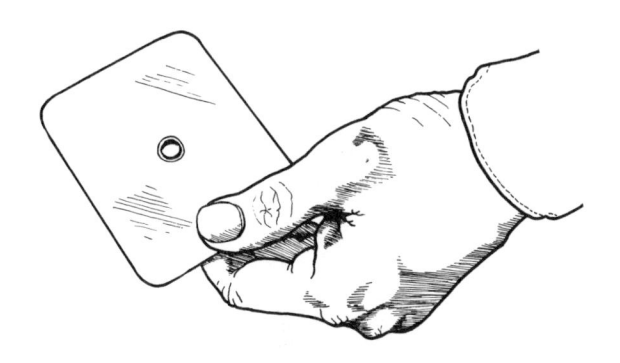

7. **Make a large sign on the ground.** Write out "HELP" or "SOS" using such materials as branches, rocks and cholla stems. Select a clear spot, and be sure to use materials that will contrast with the ground. Use jackets, emergency blankets, a fire, etc., in conjunction with your message. The message may show up better if the materials are elevated enough to cast shadows onto the ground. The larger the letters and shapes, the more visible they will be from a distance. Standard ground-to-air signals are show below. (These should be recognized by most pilots.)

I - A doctor is needed

II - Medicine is needed

X - Can NOT continue

F - Water and/or food is needed

K - Which direction do I go?

↑ - Am proceeding in direction of arrow

LL - Everything is o.k.

N - NO

Y - YES

⌐L - Message not understood

☐ - A map and/or compass is needed

8. **Carry an old sheet in your vehicle with the word "HELP" or "SOS" printed on it in large letters.** If you become stranded, raise your hood (a signal of distress); and lay your sheet out on the ground, or spread it over the roof. Don't forget to carry cord for tying the sheet.

9. **Signal flares** are potentially dangerous and can start fires, but they are often visible from great distances. We don't carry them ourselves.

In all survival situations, be sure to take care of your immediate personal needs while working on various methods of signaling. If you are unable to travel or have no idea in which direction to head, you have little choice but to try to improve your present situation. Signals are not always a sure thing. Searchers may be looking elsewhere, or your signals may not be as obvious as you think they are. Be sure to conserve your energy and your resources. Even something as large as an aircraft may take days to locate. An extreme example is an aircraft which crashed in the mountains of Arizona during World War II. It was finally discovered in 1974.

Walking Out

Elsewhere in this book, we have discussed many aspects of foot travel in the desert. If you decide that you must hike out of an emergency or survival situation, consider the following:

1. **Try to get rid of all excess weight.** Jettison any items that are not absolutely necessary to help you get out. Some items in this category might be binoculars, camera, tripod, gun, cartridges, belts, extra clothing, walking stick and other items. Every pound that you can get rid of will enable you to travel faster and farther, and even a slight difference could mean getting out or not quite making it. You can always go back later and retrieve these items. If you don't make it out, they won't have any value to you anyway.

2. **Carefully consider which route to take out. Mark it on a topo or other map if possible.** If someone is likely to come looking for you, mark your route, and leave messages as you progress.

3. **In an emergency, consider the easiest and safest route out.** You can go back for the scenery some other time. Never walk around aimlessly as this will only increase fatigue and dehydration and is not likely to get you out.

Most desert areas are laced by dry washes which make fairly good avenues. Try to remember if a particular wash crosses a highway or trail. There is usually shade along these washes, and you might come across some pockets of water. It's easier (and safer) to hike along the edges of a wash rather than in the sandy bottom. Sand is very tiring to hike in; and--at certain times--you have possible flash floods to consider. Steep, rocky washes in canyons are often feasible routes but may be very difficult to work your way through. Watch carefully for drop-offs and crumbling cliffs, especially if you must hike at night. Such a route may dead end in a steep drop that could cost you critical time and energy to work around.

CHAPTER 15 - ALWAYS

Whenever you head into the desert backcountry (whether it's for a day or a week, on foot or in a vehicle), **always leave a written message.** If you have no relative or friend with whom to leave the note, leave it with the law enforcement agency that has jurisdiction in the area where you intend to go.

The note should list the names, addresses and phone numbers of all members of your party; license numbers and descriptions of all vehicles and where they will be parked if they are to be left at a trailhead; a description and/or map of the route you intend to take; time of departure and estimated time of return. If anyone in the group suffers from a chronic illness such as diabetes or epilepsy, mention this in your note. Describe whatever medication is being taken, how often it is taken and how many days' supply is being carried. (Be sure there is medication along to see the person through an emergency or a delay.)

When estimating your time of return, allow some time for minor delays such as flat tires, taking more time to cover the trail than anticipated, etc. But be sure you specify a definite time when the holder of your note should consider you overdue and take at least preliminary steps to locate you. Don't change your plans at the last minute without advising the person who has your note. It would be a grim irony to have sent your would-be rescuers on a wild goose chase when you were in dire straits in some area far removed from the search.

Don't fail to contact the holder of your note as soon as you return so that he doesn't notify the authorities only to find out later that you have been home for hours.

The individual who can't be bothered to leave a note is often the type to land himself in a survival situation soon after making his way into the desert. When you are tempted to take off for a day hike or a backcountry drive without leaving word, picture yourself (or your vehicle) critically disabled, miles from the nearest help, running low on water, facing another day of blazing sun or a night of bitter cold and, perhaps, with friends or children depending on you to get them out. If you have left a note, how much easier it will be for you to cope with the situation until help arrives. If you have left a note and adhered to your plans, rescue is relatively certain; and you can concentrate on whatever steps must be taken to signal searchers, conserve water and protect yourself from the heat or cold until help arrives. If you have no idea when (or if) you will be missed, you may be too preoccupied with grim thoughts or moments of panic to take steps which might increase your chances of surviving and being found.

If you always leave a written message, you have already employed two of the most valuable survival techniques. Someone will be working towards you from the other end, and your mind will be clear enough to cope with the situation at your end until help arives.

CHAPTER 16 - DESERT HIKING

Hiking in the desert encompasses the same basic considerations as hiking anywhere else--plus some special considerations for such things as water, heat, salt, rugged terrain, snakes and chollas. Planning and proper physical conditioning will make any desert hike safer and more enjoyable.

The selection of desert footwear is very important. Heat alone can present severe problems, and insulation from the hot ground is vital. We have discussed problems related to fitting of boots, traction and protection from sharp rocks, cholla and other vegetation.

Always carry several pairs of socks. We often change socks at rest stops to give our feet an opportunity to cool off and dry out. Hang the sweaty socks on the back of your pack so they will be dry at the next stop. Before using them again, be sure they are free of sand and stickers. If your feet are badly swollen, you may not want to risk removing your boots (you might not be able to get them back on), but you can at least loosen the laces.

Blisters will make a trip miserable, and prevention is easier than curing. Hike often, and be sure that your boots are well fitted and completely broken in. If you can't get out often, wear your boots around home enough to keep boots and feet compatible. Be sure that they are properly treated to keep them soft and pliable and water repellent. Keep the insides clean, and never wear lumpy socks. At the first indication of a sore spot developing, stop and take your boots off. Identify the problem, and cover the tender area with a band-aid or a piece of moleskin with a hole cut in the middle so that the sore spot is surrounded but not covered.

If the day is hot, keep your hike short. Short hikes are often more rewarding than longer ones anyway. Those people who cover 20 or more miles in a day are too preoccupied with their mileage record to fully appreciate the outdoor experience. A five-mile hike in hot weather would be considered a long one under most circumstances. Limit your activities to early in the day when it's cooler.

Experience is the best teacher in the hot desert. Some years ago, we would occasionally out-hike our water supply--but not any more. Today we realize how important water is in hot climates and how quickly you can perish without it. Many people seem to require some tangible warning or a bad scare before the real meaning of what they have read strikes home.

Oftentimes the best time to hike out in an emergency situation is very early in the morning. Even before sunrise, there is usually enough light to get started, and the air and ground temperatures are usually fairly cool.

Wintertime temperatures in the desert are often ideal for hiking. The days are usually warm (occasionally hot), and the nights are cool or cold. Don't underestimate your needs for water, salts and shade in winter. While they are not as great as in hot weather, the low winter humidities often contribute to a greater need for water than you might guess. On a recent all-day hike in the Arizona desert in March, one gallon of water per person for the day proved to be just about right, even though the thermometer was only in the low 80's at the warmest time of the day. Had the temperatures been anywhere near 100° F, we are sure that each of us would have required closer to two gallons of water for the same hike. This hike covered only 12 miles and involved only moderate elevation change.

Elsewhere we have discussed some aspects of night hiking in the desert under emergency or survival conditions. In many desert areas, short nighttime hikes can be very rewarding. Be sure that everyone is wearing sturdy, high-topped boots and thick jeans and has a good flashlight with spare bulbs and batteries.

In many ways, the desert comes alive at night. Many species of desert animals are nocturnal, and many species of plants bloom only at night. One spectacular example is the night-blooming cereus of the Sonoran desert. This spindly, drab-looking plant has a spectacular flower. We once spent the better part of an hour watching the unbelievable variety of insects which visited a flowering desert shrub at night. The next day there were only a couple insects on it.

Acclimatization could be discussed under a variety of chapters in this book. It involves gradually adjusting to a particular thing. Acclimatization to the hot desert environment, if properly done, can greatly reduce the incidence of problems related to heat. The U.S. Marine Corps has all but eliminated heat-related problems by proper acclimatization of soldiers to the desert. If you live elsewhere and plan an extended desert outing (especially if it involves physical exercise such as hiking), consider arriving a week or more ahead of time to gradually accustom yourself to the heat. Start with very short outings, and gradually increase their length each day. Weight should be gradually increased also. Don't try to decrease the amount of water you need eacn day; this doesn't work.

CHAPTER 17 - VEHICLE TRAVEL

The desert presents many potential problems for the vehicle traveler. Such hazards as heat, sun, water and bites are likely to cause fewer problems than the common car accident; still, the desert poses special problems associated with rough terrain, heat, sand, high road centers, cactus and other spiny vegetation, narrow roads, long distances to help and gasoline, washes, rocks, etc. Prudent driving is required on unpaved backcountry roads. Those people who don't slow down and take a little extra care on rough, narrow roads usually pay for their stupidity in a variety of ways.

Always stay on established roads. Never cut cross-country in the desert with a motor vehicle. Off-road travel is a mindless, destructive recreation, and if you don't damage yourself or your vehicle, you will certainly damage the desert. Wagon tracks of the '49ers who passed through Death Valley more than a century ago are still visible today, as are tank tracks from World War II maneuvers in parts of the Mohave Desert.

Whenever roads are rough, allow plenty of time for the trip. Speeds of five miles per hour may be prudent in some areas; forty miles per hour may be safe enough in others. Even at five or ten miles per hour, it doesn't take long to get out of easy hiking distance to help should you become stuck or have a mechanical breakdown requiring help.

Carrying plenty of water in a vehicle doesn't present the weight problem which it does for the foot traveler. Always carry a couple of extra days' supply of water for both you and your vehicle's radiator.

Once you become mired down or stranded, you are [in many ways] in the same situation as the hiker far out in the desert. If you have plenty of water and shade, consider waiting for help if it is a long hike out and temperatures are high. We have seen statements in some survival books such as "Always stay with your vehicle." We have emphasized over and over again that things are not that simple in a survival situation. True, there are many times when staying with the vehicle is the sensible thing to do; at other times, however, it could have fatal results. The desert traveler must be able to consider all the factors and alternatives and make the best decision based on his situation at the time.

A vehicle stranded in the desert may be more visible to searchers than a lone person. Should you decide to stay put, raise the hood and tie your "HELP" or "SOS" sheet to the side or top of the vehicle. Get at least a short distance away from the vehicle as the metal will have absorbed great amounts of heat which you must avoid. Get up off the ground. Tear out a seat cushion or sit atop a spare tire. If you have to be near the vehicle, be sure that the windows are rolled down to keep it a little cooler, and tie a tarp to the shaded side.

If you decide to abandon your vehicle and hike out, leave two notes--one under the windshield wiper on the outside, the other on the dashboard inside in case the first should be blown away. Each note should give the direction you are headed, when you left, etc. In some cases, searchers have found the vehicle of a missing person soon after the search began but didn't find the body for days or even weeks, simply because they had no idea in which direction the victim was headed. Leave obvious markers as you progress. These might include arrows on the ground (made of rock or something else not likely to be washed away if it rains) or rock cairns (small piles of rocks). Broken branches may be difficult to follow but better than no markings at all. If you have plenty of paper along, impale small pieces on conspicuous branches and thorns as you pass.

Desert Driving Tips

Washes pose a particular problem for vehicle drivers. If it's obvious that a wash hasn't been crossed for some time, or if it is wet, get out and check conditions on foot before proceeding. Even a wash that appears to be dry may be mushy or slick just beneath the surface, and a car can quickly become mired down.

High centering is a common problem for drivers of passenger cars. Most two-track roads are made by trucks and consist of a high center between parallel ruts. Passenger cars simply don't have the clearance to negotiate many of these roads.

More vehicles probably get stuck while **turning around** than at any other time. Be sure to select a wide, solid spot; then make a series of short turns rather than try to make the entire turn in a single maneuver. Always keep the rear tires on the hard road surface. If you have a passenger in the car, have him get out and direct you.

Traveling with **two vehicles** is often a wise practice. One vehicle can go for parts, etc., if the other breaks down. Be sure to carry **tow chains** or tow cables and **jumper cables.**

During the monsoon season, washes that are otherwise dry can become torrents. Roads can become very slippery when wet. Sometimes a set of **tire chains** will help you get out of a slick or muddy situation, but chains are not recommended for traveling any distance.

Four-wheel-drive vehicles may have certain advantages when negotiating rough roads, but their higher initial cost and operating costs probably can't be justified by most people. Positraction and other no-slip rear ends are useful, too. **Heavy tires** (eight- or ten-ply) can reduce the number of flats due to punctures by thorny vegetation. **Wide tires** with special treads are useful for loose, sandy soils.

Carry a **come-along** (hand-cranked winch plus cable). Be sure to carry a stout metal stake or some other suitable **anchor** for the come-along because trees and other natural anchors may not be within reach. A **power winch** (either PTO or electric) can extricate a vehicle in many cases, but you must consider the added expense and weight. Have a **hook** permanently attached to each end of your vehicle so that a tow rope or chain can be attached without danger of pulling a bumper off.

Be sure that your road information is current. Check maps and make local inquiry if you're traveling into a remote area. Old mining roads which were negotiable several years ago may now be completely washed out. Some desert roads are too narrow in many places to turn a vehicle around, and backing out can be difficult. **If there is any question about the condition of a road ahead, get out and check it on foot before proceeding.**

High speeds result in considerable damage to equipment and have no place in the desert backcountry. **Speeding is one of the major causes of problems.**

Proper tire inflation is very important. Remember that pressures will be increased by higher temperatures. Driving off-road will result in many flats from chollas, mesquites and even the broken-off twigs of creosotebushes. **After driving on gravel roads, check the grooves of your tires before driving on a paved road.** Remove any potentially troublesome pebbles which may have lodged in the tread.

A heavier-weight oil is necessary to protect the car's engine in hot weather. Consider a multiple viscosity oil with SAE rating to at least 40 and maybe even 50.

Lighter-colored vehicles will be cooler inside. Avoid buying a car with dark upholstery or exterior if you live in a desert area.

Never leave pets inside a vehicle with the windows rolled up. Temperatures inside can quickly soar, and many animals have died miserably in closed vehicles.

If security is not a problem, always crack windows to avoid excessive heat build-up inside your vehicle. Cameras and film are easily damaged by heat. The glove compartment is no place to store film in warm weather.

Some Common Vehicle Problems in the Desert

This list is by no means complete yet covers many of the more common situations in which you may find yourself.

1. **Stuck:** Being stuck can have serious consequences if you are far from help and poorly prepared. As we have mentioned before, the most common way of getting stuck is in turning a vehicle around. High centering also immobilizes many vehicles. Keep an eye out for stretches of road where other drivers have come to grief. Oil spills, scratches, rubber marks and paint marks should serve to warn you of trouble ahead. Crossing washes requires special care, and you could find yourself in grave trouble if you became stranded in the path of a flash flood.

If air temperatures are high, leave your engine running while you check a situation out. If your vehicle is not bogged down too deeply, rocking it back and forth will often extricate it. Avoid gunning the motor as the spinning wheels will only dig down deeper. Tow chains or cables are a great help if two vehicles are traveling together. (Be sure that no one is in the path of a chain which might break.) Come-alongs or power winches can also be helpful. If you have a four-wheel-drive vehicle, you must be especially careful not to get stuck because a four-wheel-drive vehicle which gets stuck is often really stuck good. If you are still immobilized after attempting the above techniques, you must decide whether it is too hot to attempt more elaborate methods. If it's a hot day, it might be better to wait until evening when both air and ground temperatures are cooler. This depends on how much digging and other effort will be required.

Block the wheels. To be on the safe side, we routinely block front and rear wheels on both sides before proceeding. Two jacks are often needed. (Consider carrying one hydraulic jack and one tall, heavy-duty bumper jack.) Place a piece of plywood (carried in your vehicle for this purpose) under the jacks if you're on sand or soft ground. Carefully jack up one rear wheel at a time, and fill in underneath it with rocks, twigs and sticks or anything else which will provide traction. Build up footing under one rear wheel, lower it, then jack up the other one and build up beneath it. If the vehicle is not stuck too deeply, we have found that a long piece of carpeting placed under the rear wheels provides enough footing to get out. Don't be in too big a hurry to try the footing and attempt to get out. Do a good job of building up under the wheels the first time so that you won't have to repeat the process after the rear wheels have spun everything out the back. We once came across a four-wheel-drive vehicle with the rear wheels about four feet above solid ground. Three people had spent several hours constructing a footing under the tires and still hadn't made contact with the tires.

Once you have solid fill under both tires, carefully lower the vehicle, and check to be sure that it isn't high centered. If it is, use a long-handled shovel to dig out under it. Shovels (not just a tiny one) should be standard equipment in the vehicle of a desert traveler.

Sometimes it will help if you dig a pathway in front of the front tires. Be sure your front tires are pointed straight ahead before starting out. Try to ease out. Spinning the tires results in a loss of traction and often sinks you in deeper. It may be necessary to rock the vehicle gently back and forth. Have any able passengers help push from the sides.

Letting air out of the tires can help increase traction in some situations, but only attempt this if you have a method of reinflating them. Don't deflate too much; the tires could slip off the rims. Using an old-fashioned pump to reinflate tires is difficult and tiresome. Spark plug-type pumps are better but are somewhat slow. Consider carrying canisters which hold enough air to quickly inflate a tire. Some of these also contain sealants for temporary repair of leaks. Carry an air guage; an improperly inflated tire can cause many problems.

2. **Flat tires:** Flats can be a serious matter in many desert areas because of long distances to help and the increased likelihood of having more than one flat because of numerous sharp rocks, spines and twigs. Sometimes soft sand makes it almost impossible to jack a vehicle up unless you're prepared for the situation. Always check tires and pressures before heading out, and be sure to carry two good, mounted spares. Also carry a good pump and repair kit. Heavy-duty tires help prevent many problems, including flats.

3. **Vapor lock:** If your engine has been hot and will turn over but won't start after it has been stopped for a while, consider vapor lock as a possible cause. Wait half an hour for the engine to cool, and it will usually start right up. Wrapping the fuel pump and fuel line between the pump and carburetor with a damp cloth often helps the cooling process.

4. **Running out of gas:** Remote desert roads are no place to run out of gas. It may be days or even weeks before another vehicle passes your way. Distances between gas stations often exceed the capacity of the average gas tank. Consider having an auxilliary tank added to your vehicle, but remember that the additional weight of the tank will also reduce your overall fuel economy. Carry an extra five gallons of gasoline in a metal container affixed to the outside of the vehicle. (Never carry gas inside your vehicle.)

5. **Damaged tie-rods:** Because of their exposed location, tie rods are easily bent by oversized rocks. If you hit a rock, get out and check your tie-rods. You can sometimes straighten them enough to get you out by using pressure from a hydraulic jack. Sometimes the rods have to be reversed before they can be straightened out, then put back in the right position. Hammering can also

straighten them but must be done carefully. Such repairs should be considered makeshift, and permanent repairs should be made as soon as possible.

6. **Punctured fuel tank:** Emergency repairs can often be made using a sheet-metal screw to hold a rubber washer next to the tank, backed by a metal washer. Carry an assortment of screws and washers. We have made a plug for a larger puncture using a carved piece of soft wood with a piece of cloth wrapped around it.

7. **Hole in oil pan:** Only under the most unusual circumstances would we attempt an emergency repair of a hole in the oil pan because of the danger of losing the engine. Screws and washers might work in some situations. Be sure to carry plenty of extra oil. In such a case, we would normally prefer to hitch a ride in, get the parts and plenty of oil and make repairs in the field.

8. **Shocks and springs:** Good, heavy-duty shocks are needed in rough desert terrain. Routinely check them along with other maintenance items. Reduced driving speed will reduce the likelihood of problems.

9. **Cooling system problems:** Problems associated with the radiator are very common in the desert. Along some steep, hot stretches of highway, it is not unusual to see several drivers pulled off, waiting for their engines to cool. Many problems can be eliminated by turning off accessories such as air conditioners if the engine is running hot. Some highways have barrels of water marked "FOR RADIATOR USE ONLY" placed along steep hills.

Before hot weather begins, have your radiator cleaned and back flushed, and add the proper amounts of summer coolants. Have a radiator overflow container added. They only cost a couple of dollars but will help save antifreeze that would normally flow onto the ground. Wash out the spaces in the radiator using one of the power sprayers at a car wash. This should be done often. Be sure that nothing is in front of the radiator to retard air flow. Place the spare tire rack on your pickup off to one side so that it will not block air. When ordering a new vehicle, be sure to get an oversized radiator if you anticipate a lot of desert driving. Check to be sure that you have the correct thermostat for summer driving. We don't have an air conditioner in our vehicle and wouldn't have one. They contribute too much to overheating problems and complicate many maintenance items.

If your engine is overheating (guages are much better indicators than the so-called idiot lights), pull off the road, turn the front of your vehicle into the wind, and keep the engine running. Leave the hood up to signal for asisstance and to provide additional air circulation for cooling. Don't take the radiator cap off until the engine has completely cooled. Slowly pour water over the outside of the radiator (but not on the engine). If the temperature doesn't come down, you will have to stop the engine and wait for cooler temperatures before proceeding.

Leaks in hoses are best taken care of by replacing the hose, although special tape is available for making emergency repairs. Needle-nosed pliers can be handy for pinching off small leaks in the radiator. Commercial compounds are also available to add to radiators to stop small leaks. We routinely carry spare hoses, clamps, belts and a water pump.

10. **Brake lines:** If you break or puncture a brake line, the tube can be pinched off with a pair of pliers. Be sure to check your brake fluid level and add fluid if necessary. Carefully check your braking ability as you may have only partial braking power.

11. **Other problems:** We have covered some of the more common

mechanical problems to be encountered in the desert. There are many others. Makeshift repairs, if you have an assortment of materials and a good tool chest, will often get you home or into a garage. Keep your vehicle in excellent shape. What might be a minor problem in the city could be the prelude to a disaster if it happened in a remote desert area.

Vehicle Equipment and Supplies

It is easy to overburden your vehicle with gadgets and gear that you will never use. We see many people who go camping and spend most of their weekends tinkering with generators, washing dishes and otherwise tied down to everyday chores. They are always fussing over some piece of equipment and don't have time to enjoy the country they supposedly came to see. Travel light--yet don't leave anything behind which is necessary or which would be sorely missed in an emergency.

The following is the checklist which we go over before heading out. It doesn't include items such as food. Add to it those items which you find personally useful, and make deletions wherever you can.

Axe (or hatchet)
Bag (sleeping - one per person)
Belts (fan)
Blankets (cloth; emergency)
Bolts (variety of sizes and types)
Cables (jumper; tow)
Cap (distributor)
Carpet (1' X 5' remnants)
Chains (tire; tow)
Clamps (hose)
Coat (rain)
Coil (electrical)
Come-along (with anchors)
Compass
Condensor
Containers (gasoline; water)
Coolant (radiator)
Coolers (water; ice)
Extinguisher (fire)
Filter (oil)
Flares
Flashlights (one per person; large one for vehicle)
Fluid (brake)
Food (canned, three-day supply)
Fuel (cookstove)
Fuses
Gaskets (sheet of cut-to-size; Permatex gasket former in tube)

Gasoline (five extra gallons)
Gloves
Hammer
Hoses (heater; radiator; siphon)
Iron (tire)
Jacks (bumper; hydraulic)
Kits (emergency - see Chapter 12; first-aid; tire repair)
Knife
Light (12-volt fluorescent)
Maps
Mirror (metal signal-type)
Nuts (variety)
Oil (motor)
Opener (can)
Papers (note; toilet)
Pencils
Plugs (spark)
Plywood (heavy enough to pad jacks)
Points
Pumps (fuel; tire; water)
Radiator (commercial leak-stopper)
Rope (tow)
Saw
Screws (variety, sheet-metal)
Sheet (old bed-type with "HELP" or "SOS" on it)

Shovel (long-handled)
Soap
Spotlight (portable)
Springs (variety)
Stove (cook)
Sunglasses
Tapes (electrical; hose repair)
Tarp (plastic)

Tires (two mounted spares)
Towels (cloth; paper)
Utensils (cooking)
Washers (variety of sizes; metal; rubber)
Water (ten gallons minimum)
Wires (insulated; uninsulated)
Wrench (spark plug)

CHAPTER 18 - AIRCRAFT IN THE DESERT

We aren't pilots, so we contacted the FAA for specific hints for flyers in desert areas. Their booklet, "Tips on Desert Flying," covers many aspects of the subject; and the following information is taken from it. We included this section for general interest only, and we strongly recommend that you get the FAA booklet and obtain additional information from experienced desert pilots.

One suggestion the FAA makes is that you fly over well-traveled routes wherever possible. Included here would be railroad tracks and major highways (the latter providing a potential emergency runway). Flight plans should be filed. Travel into Mexico requires special permits. Carry at least one gallon of water, preferably in several different, tough containers. Sunglasses and a hat should be included with all flights as should your emergency kit. The FAA publication lists other recommended gear as well.

The FAA warns against using full throttle for warmups in sandy areas. Not only will the sand ruin an engine but a propeller as well. Higher temperatures, as well as the high altitudes associated with "high desert" areas, will influence your plane's performance. Be sure to study the take-off and climb-performance charts for your craft under various conditions likely to be encountered in desert areas. Novices to desert flying should consider longer runs for take-offs and longer rolls for landings. Go over these things with people who have had considerable experience with desert flying.

Winds (see Chapter 8) can seriously effect your craft. Even a "dust devil" can flip a taxiing plane under certain conditions. If you can't avoid one in your path, adjust airspeed for maximum control. If at all possible, circle around and wait for the dust devil to get out of your landing path. Dead stick landings in rough desert terrain should be avoided. If you must make a forced landing, try to do it while you still have fuel.

Once a pilot is down in a remote area, he is--in many ways--in a similar situation to that of a stranded hiker. In many cases, a pilot is even farther from roads and other sources of help than a hiker. The pilot has the advantages of having filed a flight plan, having radioed in for emergency help and having a craft that will be easier than a lone individual for searchers to spot. He is likely to have supplies of water, food, emergency items and a shelter. Fuel and oil can be used for signaling purposes. There are many variables involved in surviving in desert areas with a downed craft; but, again, planning and preparation will help increase your chances of survival.

CHAPTER 19 - DESERTS IN GENERAL

Deserts cover about one fifth of the surface of the earth. In North America, the percentage of coverage is not as great, but it is still about five percent of the land mass. Deserts are usually defined as areas where the annual precipitation is ten inches or less with high temperatures for parts of the year. Other factors, such as low relative humidities, are usually characteristic of most desert areas.

The desert of North America covers some half million square miles, but the amount of unspoiled, wild desert is quickly disappearing and is considerably less than this figure. The desert areas of North America are usually subdivided into four separate desert areas based on geography, plant and animal life and other factors.

The **Great Basin Desert** covers portions of Nevada, Utah and several other states. It is often referred to as "high desert" because of its higher altitudes and the cooler temperatures usually associated with it. Sagebrush is one species of plant often associated with the Great Basin Desert.

The **Mohave Desert** lies mainly in California. On its edges, it blends in with parts of the Great Basin and Sonoran Deserts. The Mohave shares with the other deserts such plants as creosotebush and mesquite. Death Valley lies within the Mohave Desert.

When you think of a "typical" desert with saguaro cacti and other species of large cacti, you are usually thinking of the **Sonoran Desert.** Portions of Arizona, Baja California and Sonora, Mexico, are in the Sonoran Desert.

The **Chihuahuan Desert** is mainly in Mexico although small parts of it extend into Arizona, New Mexico and Texas.

Deserts are truly fascinating places and far from lifeless. Much of the animal life, however, is nocturnal so that even many lifelong residents of desert communities fail to realize how much life exists in the desert around them.

CHAPTER 20 - DESERT PLANTS

We have briefly mentioned elsewhere the possibility of using native plants as food in an emergency situation. It is our opinion that there are very few situations where the gathering of such plants would make a difference in terms of surviving or not surviving. There is always the possibility, but certainly the ease with which one is supposed to be able to go out and gather and prepare these foods has been over-simplified by many authors of "survival" books. Those authors who give the impression that the desert is a cornucopia of edible foods and potable water give us the impression that they have never set foot in the desert.

Native Plants as Food

We would like to expound a bit further on the collecting of edible native plants under emergency conditions; then you must decide for yourself. Many of these plants are unfamiliar to the digestive systems of most of us, and the possibility of adverse reactions to even non-poisonous species should be considered. There are also many cases where toxic and non-toxic species look very much alike and could easily be mistaken by the amateur. Even getting mildly sick in a hot desert survival situation could make the difference between being found alive or dead.

We have studied a number of books covering edible plants, many of which are found in desert areas. In addition, we have tested a number of edible plants, although we do not consider our own studies to be in any way comprehensive. At times (especially in late summer) we have found large quantities of edibles such as the fruit of prickly pear cacti. Overall though, we have come to several conclusions which we think you should consider, keeping in mind that there are many people who do not agree with us at all. Ours is a conservative attitude and tends to put emphasis on prevention, staying safe and not taking risks.

1. Although there are many plants and plant parts which are edible, many of these are seasonal, and crops are not consistent from one year to the next.

2. Edibles are not found everywhere. In fact, there may be areas comprised of many square miles where virtually nothing edible can be found.

3. The opinion of a botanist may be required for positive identification of some "edible" species.

4. The nutritional value of most of these desert plants has not been determined, so whether or not they would even provide you with needed energy, protein, etc., may be doubtful in some cases. A green salad wouldn't sustain you for long in a survival situation!

5. There are often many hazards associated with the collection of plants. While these may seldom be given much consideration under normal circumstances, in a survival situation they could spell disaster. Included here are unnecessary exposure to heat and wind, cliffs, loose rocks, glochids and spines of cacti, etc.

6. Some desert plants need special preparation or cooking before they are really edible.

7. Some supposedly edible desert plants may not be so at all. Jojoba nuts,

while supposedly edible, have been proven toxic when fed in quantity to laboratory mice. We have eaten small quantities without ill effects, but this plant, as well as many others, requires further study.

We do not believe that we would be doing you any great service by describing and picturing a number of edible desert plants (for the reasons listed above). If you wish to pursue this as a hobby, we suggest that you check other references and carefully consider the sources of your information. Be conservation-minded; pick only the fruit, nuts, etc., leaving the rest of the plant to provide next year's fare.

Poisonous Plants

The likelihood of being poisoned by a desert plant is not great, but the possibility does exist. Again, we suggest that you consider other sources if you wish to pursue this topic.

Poison ivy does occur in a few desert locations, usually where moisture is found.

Should you decide to eat desert plants, avoid formulas. Whether or not it tastes bad, has red beans or milky sap may be useful for eliminating some plants from your list, but there are always exceptions. We suggest the following: "**Never eat any native plant unless you are positive of its identification, are positive that it is edible and are positive that you will not react adversely to it.**" What may be palatable and digestible for one person may be very disagreeable for another.

One plant which we mention because of its abundance near homes and in many campgrounds is the non-native oleander. This plant is poisonous, and you can become very ill just from using one of its branches to spear a marshmallow or hot dog for roasting. Be sure to caution children about oleanders.

CHAPTER 21 - DESERT ANIMALS

Few topics are plagued with more misinformation and old wive's tales than the topic of desert animals. The truth is that your chances of dying from the bite or sting of a venomous desert animal are very remote. Because of the disproportionate concern expressed by many people with regard to desert animals, we will pursue this topic in greater detail than the danger really warrants.

Snakes and Snakebite

Snakes invariably head the list of maligned desert animals. There are a number of poisonous species in the desert, and some people do get bitten. A few even die, but the number is very small. Your chances of dying from a man-caused incident in the desert are many times greater than of dying from snakebite.

About ten people per year, on the average, die from rattlesnake bites in the entire United States. When you consider that many of these bites didn't even occur in desert areas, that improper first aid (perhaps none at all) may have been administered in at least some of the cases and that many of the victims were actually handling or attempting to collect the snakes when bitten, it becomes obvious that your chances of being bitten while out hiking in the desert are very small indeed.

Rattlesnakes are usually easy to identify. Many species have a large head, broadly triangular in shape. The tail usually has a number of rattles on the end (sometimes lost), and the dorsal patterns are fairly distinctive. Even non-poisonous snakes may vibrate their tails rapidly in dead leaves or dry grass, giving a good impression of a rattlesnake.

Snakes are less active in cooler weather, and some species may den up for the entire winter. Rattlesnakes don't always follow set rules, however, and you could see one almost any time. On warm nights, often after a rain, there seem to be more snakes out than at any other time. It is not unusual to see rattlers basking on the still-warm pavement of desert highways after dark. On cooler mornings, we have seen some snakes that were so sluggish they could hardly strike. But even on a day when the air temperatures are low, a snake may have absorbed enough heat from the sun or from a warm surface to make it surprisingly agile.

Most snakes will try to avoid you and will just slip off into the brush. We know of some hikers who see rattlesnakes on almost every hike they take; yet in the process of writing two hiking guides (one in Arizona and one in Texas and both covering many areas where rattlesnakes are known to occur), we didn't see a single rattler while out hiking. Certainly we must have walked near some, and we have seen plenty of them in other situations.

Rattlesnakes are often heard before they are seen, but they do not always give a warning before striking. The sounds produced by different species vary both in intensity and quality. Some sound like the faint calls of insects while others produce a characteristic buzz that is easily heard.

A number of species of rattlesnakes occur in the Southwestern deserts of the United States and in Mexico. Some may be encountered in a variety of habitats, elevations and terrains while other species are very localized in distribution. Among the most commonly found rattlesnakes are: Black-tailed Rattlesnake (Crotalus molossus); Mohave Rattlesnake (C. scutulatus); Sidewinder (C. cerastes); Speckled Rattlesnake (C. mitchelli); Tiger Rattlesnake (C. tigris); and Western Diamondback Rattlesnake (C. atrox). There are several other species which are occasionally found in desert areas plus a few more which are found in the mountainous areas near the deserts.

Rattlesnakes are all members of the family Viparidae. They are called "pit vipers" because of a special pit located between the eye and the nostril. Cottonmouths and Copperheads of the eastern United States are also pit vipers. The pit is a heat-sensitive organ used in the capture of prey at night.

Rattlesnake venom contains a number of chemical components which include anticoagulants, tissue-digesting substances, chemicals which aid in spreading the venom throughout the body, and neurotoxic chemicals. The percent and qualitative aspects of the various components differ from one species to the next. The Mohave Rattlesnake, for example, is more dangerous than most rattlers of its size because its venom contains a larger percentage of neurotoxins.

The fangs of rattlesnakes are large and located on the front of the upper jaw. They are moveable and swing outward to a more forward position when striking. In general, the larger the snake, the more venom it can inject and the greater the potential danger; but there are too many variables involved to say that one snake is more dangerous than another. Some rattlesnake bites don't even involve the injection of any venom, while a snake which has not fed for some time may inject a relatively large amount. The bite of a rattlesnake will usually leave two large fang marks plus a set of small marks from the other teeth. The bite marks of non-venomous snakes usually show only a series of smaller tooth marks.

Methods of first-aid treatment have changed over the years and have been improved upon. Be sure to be thoroughly familiar with the present methods, and carry in your first-aid kit any materials necessary for proper treatment. Many of the methods previously advocated are no longer recommended as they often resulted in more permanent damage from the first-aid treatment than from the venom itself. In fact, there have been several deaths recorded as a result of improper first-aid treatment. In some cases, no venom had even been injected; in at least one case, the snake turned out to have been non-venomous.

There are several things which all methods of first-aid treatment have in common. **The victim of a poisonous snakebite should be kept as quiet as possible;** running or getting excited will definitely cause problems. The victim should not spend time and effort trying to collect the snake as the symptoms will indicate whether or not envenomization has taken place. The services of a physician will be required. Alcohol is definitely not among the recommended first-aid treatments for snakebite.

The doctor may or may not administer antivenom. (Some doctors in the Southwest may never have seen a case of poisonous snakebite.) Most hospitals and many clinics have the antivenom on hand or can readily obtain it from nearby sources. Pain and swelling often accompany a snakebite. Some people react to the antivenom, and the physician will test for sensitivity before administering it.

Ugly wounds which require considerable time to heal may result from rattlesnake bites, but deaths from snakebites are relatively rare in the deserts of the Southwest. Whole years have gone by in Arizona without a death, and those which do occur often result when the victim panics and runs or is given improper first-aid treatment. Many of the more severe bites occur when people attempt to capture or hold rattlesnakes. As a result, they get bitten in areas where the snake is able to inject more venom than if it were striking through jeans or boots.

The Arizona Coral Snake (Micruroides euryxanthus) is a member of the family Elapidae. Included in this group are the cobras of the Old World. One seldom encounters a coral snake as it is usually a burrowing snake and seldom surfaces. Most specimens are rather small, often being less than two feet long. They are very colorful snakes but should never be collected or harmed in any way. The head has a black front, and the body has black, red and yellowish (or whitish cream) bands which usually circle the entire body. The red and black bands alternate and are separated by the yellowish bands. An old saying, "Red and black, friend of Jack; red and yellow, kill a fellow," is useful in remembering the pattern sequence but is not very accurate otherwise. There are several other species of Southwestern snakes which are predominantly red, yellow and black; but they don't usually have the sequence of color pattern that the Arizona Coral Snake has. Included are the Long-nosed Snake, Milk Snake, Sonora Mountain Kingsnake and California Mountain Kingsnake.

The "kill a fellow" part of the rhyme is not very accurate. The Arizona Coral Snake has a very small head, and the tiny fangs do not extend as do those of rattlesnakes. It would need to nip into a thin fold of skin in order to bite effectively. Even though its venom is very potent and paralytic in its actions, we have come across no records of deaths resulting from the bite of this snake.

Lizards

There are only two species of venomous lizards in the world. Both occur in desert areas. The Gila Monster (Heloderma suspectum) occurs in Arizona and in Sonora, Mexico. The Mexican Beaded Lizard (H. horridum) is limited to a few areas in Mexico. Ignorance has caused the deaths of many of these interesting lizards. Arizona now protects the Gila Monster by statute but seldom by enforcement.

The Gila Monster is about a foot long and has a heavy tail. The head is somewhat rounded and rather large. The skin is distinctive, being bead-like in texture. It is black, mottled with shades of orange or pink. Usually this lizard appears to be sluggish as it walks slowly along, occasionally sticking out its black, forked tongue. If picked up or disturbed, it can react quickly and can easily grab a wrist or a hand. It is not aggressive by nature and will leave you alone if it isn't molested. Venom is not injected into a wound but rather flows around the teeth and into it. The grip of its jaws is very tenacious.

We have discovered only a single record of death as a result of the bite of a Gila Monster, and there is some evidence that factors other than just venom were involved. Should anyone be foolish enough to get bitten, the services of a physician should definitely be sought. We have not come across any records of deaths from the bite of a Mexican Beaded Lizard, although records are seldom kept in small villages in Mexico where there may have been a few instances. Both species of lizards are similar in appearance and behavior, and their venom is probably similar as well.

Venomous reptiles found in the southwestern deserts of the United States include, from top to bottom: Sidewinder (Crotalus cerestes), **Gila Monster** (Heloderma suspectum), **and the Arizona Coral Snake** (Micruroides euryxanthus).

Scorpions

All species of scorpions are poisonous, but one species in the United States is responsible for most deaths and serious reactions. The culprit is the Bark Scorpion (Centruroides sculpturatus).

Scorpions as a group are easy to identify. They have four pairs of legs, a pair of large pedipalps with "pinchers," a cephalothorax and a segmented tail which has a bulbous end with a stinger. The tail is flexible and can easily be brought around to sting anything touching the head or the back of the scorpion. Scorpions usually eat insects.

The Bark Scorpion is a relatively small species, being about three inches long. It is slender and usually rather light in color, although one form of it has a darker pattern on top. There is a small tooth near the base of the stinger (magnification may be needed to see it clearly). There are other potentially fatal species in Mexico and may be several others in the United States which are suspect.

In Arizona in the past, scorpions (especially the Bark Scorpion) have been responsible for more deaths than rattlesnakes. Records indicate that from 1924 through 1954 there were 64 deaths from scorpion stings in Arizona but only 19 deaths from rattlesnake bites. From 1960 through 1969, there were four deaths from scorpion stings, an indication of improved medical treatment. Antivenom is available at many hospitals for the sting of Centruroides. Deaths from the stings of desert scorpions have also been reported in Mexico, California and Texas.

You should exercise caution wherever Centruroides is found. We have been stung a total of three different times, all in an old adobe house in Tombstone, Arizona. The place was crawling with scorpions, and it's amazing that we weren't stung more often. Rockhounds who use black lights at night to find minerals which fluoresce should be very cautious as many species of scorpions will also fluoresce in black light. The Bark Scorpion can easily enter cracks only 1/16 inch thick. Always check sleeping bags, shirts, boots, socks and bedding for scorpions. Wear gloves if you are going to be handling old lumber, boxes or trash. Don't wander around a house bare-footed if you know that scorpions are commonly found there. Victims of stings should be checked by a doctor. Some people have severe reactions, and individuals who are especially young, old or ill may be in danger if proper care is not sought.

Spiders

1. Black Widow Spider: Females of this species (Lactrodectus sp.) are usually very distinctive in appearance, being a shiny black or brown and having a reddish hourglass on the underside of the abdomen. Males are much different in appearance and do not present a problem. Widow webs often appear rather haphazardly constructed, and the females are often found hanging upside down in them. The egg cases are often easy to spot. In some desert areas, Black Widow Spiders are abundant. We once found 50 under an old refrigerator door which had been lying on the ground. Saguaro skeletons, old cardboard boxes, piles of lumber, the undersides of sinks, the interiors and undersides of outhouses and piles of debris--all these are likely habitats for Black Widows. Be especially careful around old buildings.

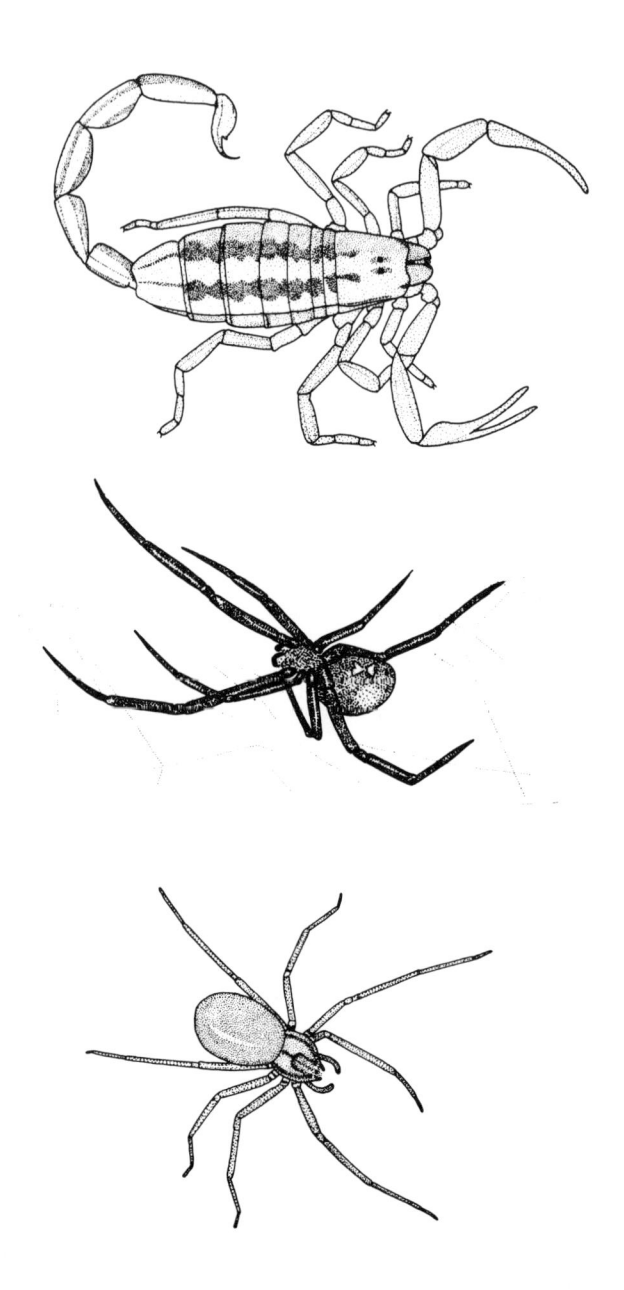

Invertebrates which are poisonous to varying degrees and are possible hazards to careless desert travelers include, from top to bottom: **Bark Scorpion** (Centruroides sculpturatus), **Black Widow Spider** (Lactrodectus sp.), **and the Recluse Spider** (Loxosceles arizonica).

People are usually bitten by Black Widow Spiders when they accidentally come into contact with them. Pain and nausea often occur as a result of the bite, although a victim may not realize at the time that he has been bitten. The venom contains a neurotoxin. Although death from the bite is rare, there are records of deaths occurring. Again, the young, the old and the ill are more susceptible. A doctor should see anyone who has been bitten by a Black Widow.

2. **Recluse Spiders:** There are several species of Recluse Spiders which occur in the desert Southwest. They are commonly referred to as "violin spiders" because of a violin-shaped pattern on the cephalothorax. There is often little or no pain at the time of the bite, and many victims don't even realize that they have been bitten until later. In a few cases, there has been a general reaction to the bite; and there are a few cases where death resulted. A small, necrotic ulcer may develop at the site of the bite and may take a long time to heal. We have usually found Recluse Spiders in old saguaro skeletons, but they may be found in a variety of situations.

Other Desert Invertebrates

There are a number of other invertebrates commonly found in the desert that are either poisonous or thought by many to be poisonous. Some of these are quite benign and interesting despite their sinister appearances.

1. **Tarantula:** These large, hairy spiders may be rather common at times. In certain locations, we have seen large numbers of them crossing highways late in summer. (It is usually the males that are seen moving about.) Their fangs are large, and they can give you a well-deserved nip if you pester them. These spiders should be considered beneficial. They may be long-lived, some specimens being more than ten years old.

2. **Honeybee:** Honeybees are common in the desert in most agricultural areas as well as in some remote canyons. They may also be found in hollow trees. In the United States, they are responsible for more deaths than all other venomous animals combined. If you are a hypersensitive individual, the sting of a Honeybee is a serious matter which requires immediate first aid. If you know you are liable to react violently, you should consider carrying prescription drugs in your emergency kit. Other people may suffer little or no reaction to these stings. Avoid squeezing the poison sac when removing it; try to scrape it free without injecting more venom into the wound.

3. **Velvet Ant:** The Velvet Ant is actually a wasp, and the wingless females are commonly seen scurrying across the desert floor. They look soft and fuzzy and come in an assortment of pretty colors (reds, oranges, yellows, golds and whites)--and they can give you a painful sting if you tamper with them.

4. **Ants, Wasps, Hornets:** There are many species of ants, wasps and hornets in the desert. Most of them bite or sting, and a few of them are able to get you with either end. Leave them alone. Some people may react adversely to the bites or stings of certain species.

5. **Mosquitoes, Gnats:** These are seasonal and are often local in occurrence--and supremely annoying when they descend on you in large numbers. Be sure that your tent is insect proof. Insect repellent usually helps to varying degrees.

6. **Giant Desert Centipede:** These may reach nearly a foot in length and

are found under logs and rocks. Their bites can be painful, but envenomization is usually mild. Formation of small ulcers at the sites of bites have been reported in a few cases. The Centipede is usually not agressive unless picked up or trod upon by a bare foot, at which time it will usually bite.

7. **Giant Millipede:** After late summer rains, you may see a number of these crossing desert highways. They usually coil up in a tight spiral if handled, and they emit a secretion which can be irritating if it gets into the eyes. A few people have reported skin irritations after picking them up. They are not poisonous, and they do not bite.

8. **Vinegaroon:** We include this species only because many people are convinced that anything so ugly must be poisonous. It can pinch, and it does secrete a vinegar-like substance (acetic acid) if irritated. Otherwise it is harmless.

9. **Solpugid:** Also known as "sun spiders," we have most often seen these at night, busily hunting around the floor in our old adobe house. They move quickly and are more startling than frightening. They lack venom glands.

10. **Jerusalem Cricket:** These unusual insects are often found under rocks. They are related to grasshoppers and other crickets. While somewhat gruesome looking, they are not poisonous, although one might pinch you if you picked it up.

11. **Cone-nose Bug:** These insects are commonly found in wood rat dens. Campers may accidentally come into close contact with them by camping near locations where they are common. They suck blood and are a potential vector of diseases, but they don't seem to be responsible for many illnesses in the desert Southwest. In South America, related species are guilty of transmitting disease.

Precautions

Listed below are some basic precautions which all desert travelers should consider in order to avoid being bitten or stung by poisonous animals. Even the most cautious individual will occasionally encounter a venomous animal.

1. **Be particularly careful where you step,** especially in areas where visibility is limited. Don't reach down to pick up a mineral specimen or other object without first scanning the immediate area for a coiled rattlesnake. Always watch ahead of you on the trail and along its edges. Climbers should avoid reaching blindly onto a ledge where a rattler may be sunning.

2. **Always carry a flashlight at night.** While falls are the most common hazard when hiking at night, venomous animals are also active after dark. Rattlesnakes usually retaliate when stepped upon.

3. **Learn to recognize the typical buzz of a rattlesnake, and be alert for it as you hike.** If you hear a rattler buzz, pinpoint its location, then give the area a wide berth. That's all he wants you to do.

4. A good many rattlesnake bites occur when people attempt to capture or handle them. There are even records of collectors having been bitten through the thick specimen bags which are used for transporting snakes. **Leave all snakes alone.** Observe them from a safe distance through binoculars. (Most species can be readily identified using this technique.) Should you decide to photograph a rattlesnake, use a long telephoto lens, and have someone else keep an eye on the snake while you are preoccupied with camera adjustments. Even a supposedly "dead" rattlesnake can still have a functioning nervous and

muscular system. There is at least one record of a man dying after being bitten by the severed head of a rattlesnake. Rattlesnakes give live birth, and even the tiny ones are very venomous and should be left strictly alone.

5. **Be especially cautious when working around rock, brush, wood piles, logs, saguaro skeletons and various types of desert debris.** Rattlesnake bites commonly occur in areas of high grass. If you step over a rock or a log in your path, be sure to visually check the other side for a rattlesnake. If you must pick up a rock or log, tilt it back and up towards you so that a coiled snake would not be able to strike directly at you. The undersides of rocks, logs, cardboard, lumber, etc., are also good scorpion habitats; and you should make it an inflexible rule never to place your fingers beneath a surface where you can't see.

6. **Wear high, heavy boots and heavy denim jeans in snake country.** Snakes are capable of striking through these, but the probability of envenomization is reduced. People who work outside all day in areas where visibility around their feet is limited (telephone repairmen, for example) should wear a pair of commercial snake guards for additional protection.

7. Scorpions (and a few other venomous invertebrates) have a habit of crawling into boots, bedding, shirts, socks, tents and other artificial habitats during the night. **Always check carefully for the presence of scorpions.** Avoid walking barefoot around camp or in an old building on a warm evening. Scorpions often blend in with their surroundings so that you are not aware of their presence until you get stung.

8. **Select a good, durable tent with a sewn-in floor and tight-fitting mosquito netting.** Not only will this help to keep scorpions out, but it will deter bothersome mosquitoes and gnats.

9. **Don't collect, tease or irritate any desert animals.** Most of them are preoccupied with their own affairs and will go out of the way to avoid trouble. Many people are tempted to "save" a young animal (especially a more appealing species) that appears to be lost. Resist the temptation. Chances are very good that (1) it is not lost; (2) it can fend for itself; and (3) it will die during the process of being "saved." Most problems occur during the process of collection. All these animals--the gruesome, the appealing, the poisonous and the defenseless--are vital components of the desert ecosystem and should be left undisturbed for others to enjoy.

EPILOGUE - THE DESERT MUST SURVIVE, TOO

Our beautiful, wild deserts are taking a beating from which they may never recover. Our government agencies, senators, congressmen, and state and local officials are often unaware or unconcerned. A well-written letter, accompanied by documentation (photographs, place names and dates) will help focus their attention on the damage being done by off-road vehicles, mining and development. The destruction of the deserts south of the border in Mexico is taking place at an apalling rate as Mexico joins in the mad scramble for money and materials without regard for long-range effects on the land, its plants, its animals--and its people.

The most needless destruction is being done by off-road vehicles in the name of recreation. Big business is making millions of dollars off motorcycles, so-called trail bikes, and four-wheel-drive vehicles. It has no interest in the desert except as a giant playground from which it derives money. The grinding and whining of these machines as they probe the canyons and barrel across the flats is sickening to anyone who has an interest in the desert in its natural state; to anyone who would gladly walk ten miles for the privilege of sharing a desert canyon with a mule deer, a cottontail rabbit and a canyon wren; to anyone who counts sunsets and sycamores among his blessings. Motor bikes have not been called "Hirohito's Revenge" for nothing. Delicate vegetation is crushed, animals are sent diving for cover--and erosion sets into the tracks to create gouges in the landscape that will be an eyesore for future generations.

We have a four-wheel-drive truck--and it stays on the road at all times. All motorized vehicles must be strictly prohibited from leaving established roadways (of which there are plenty). Enforcement, both in the field and in the courts, must be strict. Any areas set aside for off-road vehicles should be considered an ecological disaster area and biologically dead.

Many of us doubt the existence of enough time and intellect to stop man's lemming-like rush towards the brink. But we can work together to slow it down. If man is to survive, the desert (and the mountains and the seas and the trees and the animals) must survive, too.